PROSE
AND
CONS

PROSE AND CONS

The Do's and Don'ts of Technical
and Business Writing

CAROL M. BARNUM, Ph.D.

**National
Publishers**
of the Black Hills, Inc.
Elmsford, New York

ISBN: 0-935920-29-3
Cover and Interior Design: Hudson River Studio
Typesetting: The Sant Bani Press
Editing and Production: Ellen Schneid Coleman
Printed in the United States of America

TABLE OF CONTENTS

Acknowledgments

Introduction

CHAPTER 1 Getting Started   1

Brainstorming   2
Outlining   2
Reference Guides   5

CHAPTER 2 How Do You Like My Style?   7

Sentence Length and Readability   7
Paragraphs   9
Bureaucratese, Officialese, Gobbledygook, and Legalese   11
Neologisms   17
Jargon   19
Mood and Voice   20
Personal Pronouns   21
Analogy   22
Company Style Manuals   23

CHAPTER 3 Know Your Audience   25

Picture Your Reader   25
Primary Readers   26
Secondary Readers   26
Prepare To Meet Your Audience   27
Readers Come In All Shapes   27

CHAPTER 4 Let's Get Organized   31

Order the Parts   32
What Readers Read   33
Head 'em Up with Headings   33
The Numbers Game   36
Bullets—Shoot Straight   36
Numbered Lists   37
The Style of Lists   38
What's the Forecast?   40
The Summary—The Heart of the Matter   41
How To Write a Summary   44

CHAPTER 5   Graphics are Grabbers   47

Think Graphically   47
A Picture Is Worth a Thousand Words   48
Placement Is Important   48
Label and Number   49
Show and Tell   49

CHAPTER 6   Editing: The Garbage Collector   57

The Cooling–Off Period   57
Tips for the Task   58
The Sargasso Sea—An Editing Challenge   65

CHAPTER 7   Cases in Point   69

Those Who Should Know Better   69
Those Who Do Know Better   70
Those Who Know Not What They Do   74
Those Who Cannot Be Saved   77

CHAPTER 8   The Proof of the Writing Is in the Reading   81

Proofreading Pointers   82
Quotable Misquotes   87

CHAPTER 9   Diction's Dirty Dozen   89

1. Affect–Effect   89
2. Among–Between   90
3. Amount–Number   90
4. Bad–Badly   91
5. Comprise–Consist   91
6. Due to–Because of   92
7. Farther–Further   92
8. Kind–Kinds   93
9. Lead–Led   94
10. Lie–Lay   94
11. Lose–Loose   95
12. Principle–Principal   95
13. Stationary–Stationery   96

CHAPTER 10   Punctuation Pointers   97

A Few Ground Rules   97
The Comma   98
The Semicolon   102
The Colon   103
Dashes, Parentheses, Square Brackets   104
Quotation Marks   106

CHAPTER 11   The Grammar Grind   107

Subject-Verb Agreement   107
Inverted Word Order   108
Indefinite Pronouns   109
Pronoun Agreement   111
Pronoun Reference   111
Who and Whom   112
Reflexive and Intensive Pronouns   113
Possessives   114

Suggested R and R (Reading and Reference)   117

Exercises   119

Answers to Exercises   141

Index   153

ACKNOWLEDGMENTS

Some people have always known that they wanted to write a book. I was not one of these. Then one day, not long after I had joined the faculty of Southern Tech, my boss Robert Fischer said, "When are you going to write a book on technical writing?" Hmm. I hadn't thought about it . . . but he planted the seed of an idea. Still, it was going to be someday. Later.

About two years later, I got a letter from an acquisitions editor of a large publishing house, who said that she had read an article I had published in *BYTE* magazine, liked my style, and thought I ought to write a book on technical writing. Thinking that maybe someday was today, I said I would.

I was supposed to get back to this editor after I'd written a sample chapter or two. Instead, I wrote the whole book. So it was a year later when I got back to her. Meanwhile, her company's publishing plans had changed.

Now I had a book and no publisher. That problem was remedied by National Publishers. And now *you* have this book.

As there are always many more people than the author making contributions to any worthwhile project, I'd like to acknowledge some who have been particularly helpful to me: (1) Robert Fischer, for starting me thinking about this book, (2) the unnamed acquisitions editor for believing I should and could write a book about technical writing, and (3) my editor at National, Ellen Schneid Coleman, for liking the book and wanting to publish it.

I'd also like to thank the nimble-fingered Judy Waits for typing the manuscript with lightning speed, the clever and creative John Tumlin for preparing the exercises that so beautifully capture the tone of the book, and the meticulous and thorough James Carroll for preparing the index.

Finally, I'd like to thank Jack Aiken who, without even reading past the first page of this book, knew I could do it all along.

INTRODUCTION

This book is not intended as the last word on the subject of technical or business writing. The last word will probably never be written, since language is constantly changing and evolving to meet the needs of its users.

Rather, this book is intended as a beginning, a useful starting place for those who write on the job as well as for students who want to develop a prose style that's short and sweet. It is not intended as a compendium of learning, but as an adjunct to learning and a handy reference guide for those who want something they can grab from the shelf without breaking an arm.

Its format is straightforward. It presents the key points of good technical and business writing, including such burning issues as sentence length, diction, jargon, style, summaries, audience analysis, methods of organization, brainstorming, outlining, editing, proofreading, and a smattering of the worst offenders in grammar and punctuation. Its intent is to serve up such fare in digestible bites—to provide food for thought, as it were—without stuffing you to the choking point. In so doing, it avoids the jargon of English teachers as much as it urges you to avoid the jargon of your profession.

Its aim is to simplify, not stupefy.

1
Getting Started

Before you begin any project, it's wise to have a plan. But how to plan? Your boss wants a report on the project you've been working on, and since you know more about it than anyone else—you've lived and breathed it for all this time—it should be easy to write something. Right?

Wrong.

What is it that strikes fear in the hearts of most people when they are told they have to write? Many things, but you can probably identify with some or all of the following:

1. You don't feel confident of your ability to express yourself.

2. You don't write often enough to get past the writing jitters.

3. You're not sure about your command of grammar and spelling (after all, you're *not* an English major).

4. English wasn't your best or favorite subject in school.

5. Technical people aren't *supposed* to have to write.

If I told you that *no one* likes to write—not even professional writers who earn their livelihood by the pen (or typewriter or word processor)—you probably wouldn't believe me. If I told you that I have dreaded the thought of sitting down to write this book—putting every excuse imaginable in front of it, even preferring to clean out old files (not to mention the refrigerator!) than to start writing this book—you probably wouldn't believe that either. And even if you did believe me, it probably wouldn't make you feel any better about your own ability to write. Yet you know the writing has to be done and that, if you want to get ahead in your field, one key to your success is your ability to communicate what you know to others and what you want others to know.

But before you can do this effectively, you've got to have a plan, and a good way to develop that plan is through brainstorming.

Brainstorming

As a common practice among executives, brainstorming works something like this. A group of executives gets together in a conference room with a blackboard and plenty of coffee. They start a freewheeling discussion of a problem. *Any* idea, no matter how seemingly ridiculous or remote, gets recorded on the blackboard until everyone runs dry. Then everybody ponders what's been generated, trying to see if a picture emerges—some pattern, some common group of ideas, or even something no one has considered until that moment. One thought leads to another, and eventually a plan develops.

You don't need to be a business executive to use this technique. It works well for any writing project, no matter what the size. When a writing project comes up—whether someone has asked for it or you know you need to write something—you can start the ideas flowing by brainstorming. First you take a clean pad and write down every idea on the subject that comes to mind. Then you look at what you've written. Are any patterns emerging? Do you see any correlations between ideas? Are some ideas more important than others? If so, indicate the most important ones with a number or an asterisk, and rank what's left accordingly. You should now see a plan emerging for your memo, letter, or report. From that, you can conceive an outline.

Outlining

Remember the outline format you learned in school? You probably tried to forget it: Roman numerals for main points (I, II, III, etc.), capital letters for subpoints (A, B, C), Arabic numerals for points below that (1, 2, 3), and lowercase letters for points below that (a, b, c). Your teacher probably required either a sentence outline or a topic outline. In the sentence outline, every entry had to be a full sentence, and in the topic outline, every entry had to be a complete phrase. In both cases, all entries of the same level had to be constructed in the same way (see parallel structure pp. 38, 64–65. That is, *all* the Roman numerals had to be expressed in the same grammatical construction, *all* the A, B, C, D entries under any Roman numeral had to be parallel, and *all* the 1, 2, 3 entries had to be parallel with each other (although any group of 1, 2, 3 or A, B, C only needed its sentences or phrases to be parallel with each other, not with those of any other group). And finally, any A had to have a B and any 1 had to have a 2 because if you wanted to subdivide something, you had to have at least two points or it wasn't worth subdividing.

What an awkward mechanism! Besides, who really thinks in outlines? As a survey of technical writers shows, not many. Ninety-five percent of the respondents indicated that they use some form of informal outlining technique, but only five percent use a formal sentence outline.[1] If you've been forced to generate a formal outline, can you really follow it through the writing stages of the project? If you can, great; formal outlines must work for you.

Although the traditional outline format teaches good organizational skills, the mind moves in mysterious ways, most of which don't conform to the rigidity of an outline. As a result, most people give up formal outlines as soon as they are no longer required.

Still, you need some sort of outline if the writing is going to move with any logical progression from point to point. The best system is your own system, the one you devise to organize your thoughts. In my system, I move from brainstorming to organizing my ideas into groups. From there, I start writing, letting my thoughts flow as they will. It's a kind of stream-of-consciousness writing, but it's guided by the general plan I've mapped out in my brainstorming and outlining session.

My outline for this book, for instance, looked like the one on page 4.

It wasn't neat. I wrote it on a sheet of notebook paper. I didn't end up putting everything that was in the outline into the book. And I put a lot of things in the book that weren't in the outline.

But it was my starting place. Each day before I began writing, I referred to my outline to get my bearings. Such a system works for me, but it begins, even before brainstorming, with lots of agonizing over the piece that I need to write, followed by lots of thinking about it, and ending with lots of editing.

Other ways of organizing are just as good. Take the cut-and-paste method, for instance. Those who use this method know what they want to cover but aren't sure of the order, so they write a section here and a section there,

---

[1] Blaine K. McKee, "Do Writers Use an Outline When They Write?" *Technical Communication,* First Quarter (1972) pp. 10–13.

sentence length
diction
grammar
punctuation
audience analysis
outlining
brainstorming
headings
summaries
jargon
dictionaries - thesauruses
editing
spelling
word processors
methods of organization
proofreading - reading backwards
     get someone else to read
                              Audience - Editing
1. brainstorm   4. edit
2. outline   5. proofread
3. rough draft

Things your teacher told you were wrong:

Never use I
Never end a sentence with a preposition
Never a 1-sentence paragraph

then cut and paste it all together in the proper order. Still others sketch out a plan with lines emanating from a central point to the points they want to cover. A recent article describes the effectiveness of this technique.[2] The author explains her use of three visual patterns—the S-Curve, (S), the Hub and Spokes, (☿), and the Pyramid (▲ ▼)—as organizational aids for different writing situations. Sure enough, her technical students were able to organize better when they could *see* a pattern in their thoughts.

Another more traditional technique—using note cards—is helpful for many. With note cards, you can put ideas down at random and then shuffle the cards to place similar ideas together for the first draft. When you add topic headings to each note card, the shuffling and organizing process is made even easier. Whatever method you use, the point is that you've got to organize your thoughts before you begin to write, and the system has got to work for you.

With your brainstorming over and your outline before you, you're almost ready to go. But you still need one or two more things within easy reach: a good dictionary and a good handbook of grammar.

Reference Guides

Indispensable to any writer is a good dictionary. It should be big enough to contain information on uses of a word in a particular context: alternate spellings; aberrant spellings in the past tense, the plural, etc.; and other helpful points of reference. While it should be large enough to contain this kind of information, it should still be small enough to be handy. To that end, it should be kept close by, and it should be light enough to find its way into your hands *often*. No one is born a good speller, but everyone can learn to use a dictionary. The good writers separate themselves from the bad by being willing (if not eager) to check the dictionary whenever they have doubts about the meaning or spelling of a word.

While we're on the subject of reference guides, what about a thesaurus? My advice is to skip buying one unless you can trust yourself not to use it too often. The reason? Most people operate under the mistaken notion that their everyday vocabulary is OK for a talk with Joe at the water fountain but that they need to "upgrade" it when they write. Although they'd be right in recognizing that colloquial speech (the shorthand of daily conversation) is not appropriate for the formal English prose which most reports require, they generally take that thought one disastrous step too far, assuming that

---

[2] Susan Feinberg, "Visual Patterns: An Experiment with Technically Oriented Writers," *Technical Communication,* First Quarter (1984) pp. 20–21.

their everyday vocabulary needs "dressing up." That's where they bring the thesaurus in, usually to the detriment of their report.

Let me demonstrate by example. If a person wants to use another verb in place of the verb *desire,* as in the sentence, "He desires a change in jobs," he or she could find such synonyms as *ache, covet, crave, hanker, hone, long, pant, yearn for,* and *wish* in the thesaurus. By substituting any one of these words for *desire,* the writer changes the meaning of the sentence. For instance, the sentences "He pants for a change in jobs" or "He covets a change in jobs" or "He craves a change in jobs" do not convey the same meaning. If the writer knows all the synonyms for the word he or she looks up and understands the differences in meaning, then it's possible to substitute one for another in certain instances. But if the writer doesn't know all the synonyms and assumes that because they're listed as synonyms one can be substituted for another, then it is highly likely that the writer will be conveying a meaning he or she doesn't intend, not to mention the fact that there is nothing wrong with the original word *desire.*

To take the example one step further, if the writer used the word *desire* as a noun, as in the sentence, "He had a strong desire to change jobs," then he or she would find the following synonyms in the thesaurus: *aphrodisia, concupiscence, passion, prurience, appetence, appetite, craving, hunger, itch, longing, lust, yearning, yen.* Try substituting one of these for *desire* in the original sentence, and you'll see how the meaning changes.

So much for the thesaurus. It's handy when you've used the same word over and over and would like to break the monotony with a different word. Otherwise, it can be dangerous.

A reference that can be very useful, however, and one which you should have when writing, is a handbook of grammar. Handbooks come in all shapes and sizes: the only rule here is to pick one that's fairly current and that you can use easily, so that if you need to find the rule about the placement of a comma, for instance, you know where to look. This book contains the most important rules, but a good handbook of grammar will give all the rules.

2
How Do You Like My Style?

E verything has a style of its own. One year wide lapels on suits are in; the next year narrow lapels are the rage. Skirts can be long or short. Cars can be classic or futuristic. The style of technical and business writing changes, too (although not as rapidly as the style of clothing or cars). Nonetheless, certain characteristics continue to distinguish technical and business writing from other kinds of writing. Sentences are generally short and to the point. Paragraphs are also short and grouped under headings and subheadings. Many paragraphs contain lists. The words chosen for sentences are simple, clear, and direct. Although technical or business writing need not be dull, its purpose is not to entertain but to inform, persuade, or otherwise impart information to an intended audience with an express desire to know and understand a particular subject. Therefore, the style of technical and business writing is designed to get the message across clearly and quickly.

Sentence Length and Readability

Hemingway is known for his exceptionally short sentences; Faulkner, for his exceedingly long ones. Technical and business writing style falls somewhere in between these two extremes.

Much has been written about sentence length in technical and business writing and its relationship to readability. Rudolph Flesch and Robert Gunning were among the first to devise readability formulas to measure the reading ease of a piece of writing. Gunning's formula, called the "Fog Index," works like this. You take a passage of about 100 words, count the number of big words, count the number of complete thoughts, do a little multiplying and dividing, and come up with a number that roughly equates reading level with grade level. Any number above twelve (corresponding to twelfth grade level) is too high, indicating that the writing is probably unnecessarily complicated for most audiences.

Here, specifically, is how the Gunning Fog Index works:

1. Choose a passage of approximately 100 words. Divide the total number of words in the passage by the number of complete thoughts. A word is anything with space around it. (Example: January 5, 1985 is three words.) This gives you the *average sentence length.* A complete thought could be a sentence or it could be a part of a sentence containing a complete thought which is connected to another complete thought by *and* or by punctuation like a semicolon.

2. Divide the number of words of three or more syllables by the number of words in the passage to get the *percentage of hard words.* Do not count word combinations made of simple parts (ex. bookkeeper, stockholder). Also do not count verbs that get a third syllable by adding *-ing, -ed,* or *-es.* (ex. describing, created, processes). Do not count proper nouns (capitalized words like company names) since these cannot be changed.

3. Add the average number of words in a sentence to the percentage of hard words. Then multiply by 0.4. The number you get is the Fog Index.

I tried the formula against one of the passages in this book to see how I measured up. If you want to follow along, my passage begins at the start of this chapter with the word *Everything* and ends nine lines later with the word *direct.* It contains 102 words. Applying the Gunning formula, I count eleven complete thoughts:

$$\frac{102}{11} = 9 \text{ words average sentence length}$$

I count 12 "hard" words of three or more syllables. (I did not count *everything, nonetheless,* or *subheading* since these are combinations of simple words.)

$$\frac{12}{102} = 12 \text{ \% of difficult words}$$

When I add 9 + 12 = 21 and multiply by 0.4, I get 8.4 which means grade level eight. Now, try it against your own writing. How did you measure up?

While readability formulas are interesting and useful, they are not foolproof, since they do not measure the clarity of a piece of writing. For example, you could write a 100-word passage full of gibberish with all the gibberish in sentences of fewer than ten words and words of fewer than three syllables. The readability quotient would be very good even though the passage made no sense. Nevertheless, applying one of the readability formulas to your work can give you a clue about your writing style, pointing up a tendency toward excessively long sentences or overly complex words.

Paragraphs

The main rule to consider when writing paragraphs is that sentences should combine to form units of thought. But how do you know where one thought ends and another begins since all thoughts relate to the same subject? A helpful rule is that a page of double-spaced typed text should contain at least two, preferably three, paragraphs. Since a page holds 200 to 250 words, paragraphs should average less than one hundred words. The main reason for keeping paragraphs short is the need for white space, those restful breaks for the reader. Even paragraphs on the same thought should be broken for the brief respite white space provides and then connected to the next paragraph with a transition word or expression. Transitions can and should be used within paragraphs and between paragraphs.

As the following list shows, transition words and expressions can link similar ideas, set up contrasts between ideas, show chronological arrangement, spatial order, step-by-step order, results, or conclusions.

1. *For Similarity*—moreover, in addition, further, furthermore, likewise, and, also, equally important, similarly

2. *For Differences*—on the other hand, conversely, yet, but, however, still, nonetheless, nevertheless, on the contrary, in contrast, at the same time

3. *For Step-by-Step Order*—first, next, then, afterward, following, soon, in the meantime, after that, afterwards, simultaneously, later, earlier

4. *For Spatial Order*—here, there, beyond, nearby, opposite, under, over, inside, outside, above, to the left, to the right, in the distance

5. *For Conclusions*—in conclusion, finally, to summarize, on the whole, in brief

6. *For Results*—therefore, thus, hence, consequently, as a result, accordingly

You can use these words and phrases by themselves or link them to other words from previous sentences or the preceding paragraph to connect ideas from sentence to sentence or paragraph to paragraph.

Whole paragraphs can also serve as transition to new topics. These are usually called forecasting statements (see pp. 40–41) and are generally shorter than the average paragraph. They tell the reader what to expect next.

PARAGRAPH DEVELOPMENT

Paragraphs are generally developed in one of two ways—through inductive order or through deductive order. Using inductive order, the writer makes a statement, then another, then another, and the accumulation of the statements leads to a general principle or conclusion which is revealed at the end of the paragraph. Inductive order is most often used in scientific writing for developing a hypothesis or theory on the basis of a number of premises. It is best used in technical and business writing for persuasive arguments. When you don't want to show your hand at first (and there are many good reasons why you might not), inductive order allows you to present and build your case premise by premise, leading to the conclusion for which you've been preparing the reader.

In most business and technical writing, however, deductive order works best. In this arrangement, the paragraph begins with a generalization and the sentences which follow support that generalization with examples and other supporting material. The generalization is called the "topic sentence" because it announces the general topic of the paragraph. This approach works best in most situations because it allows the readers to get the gist of the paragraph at the outset. The busy reader could conceivably read only the topic sentences of a report and understand the contents. Many do just that. Since readers have come to expect deductive order, most reports should be written this way to deliver the message quickly. Not only should paragraphs be developed this way, but also whole reports, with the most important statements placed up front and followed by the supporting material. More about that later.

Bureaucratese, Officialese, Gobbledygook, and Legalese

When the government writes in its own tongue (so to speak), it produces a strange hybrid called bureaucratese or officialese. When business people use their own language, it's called gobbledygook. When lawyers do the same thing, it's called legalese. Many of the rest of us feel that such professional groups use their "native tongue" because they don't want us to know what they're saying or doing or because they don't want us to know what they're *not* saying or doing or because they want us to be "impressed" with their vast powers of communication. What they fail to remember (or remember all too well) is that their writing doesn't communicate if it *sounds* intelligent but conveys no meaning.

A playful example of the official style of the government points up the problem:

A Bureaucrat's Guide to Chocolate Chip Cookies

For those government employees and bureaucrats who have problems with standard recipes, here's one that should make the grade—a classic version of the chocolate chip cookie translated for easy reading.[3]

Total Lead Time: 35 minutes.

Inputs:
1 cup packed brown sugar
½ cup granulated sugar
½ cup softened butter
½ cup shortening
2 eggs
1½ teaspoons vanilla
2½ cups all-purpose flour
1 teaspoon baking soda
½ teaspoon salt
12-ounce package semi-sweet chocolate pieces
1 cup chopped walnuts or pecans

Guidance:

After procurement actions, decontainerize inputs. Perform measurement tasks on a case-by-case basis. In a mixing type

---

[3] Susan E. Russ, "A Bureaucrats Guide to Chocolate Chip Cookies," *Washington Post,* n.d.; reprinted in *Simply Stated,* April 1982.

bowl, impact heavily on brown sugar, granulated sugar, softened butter and shortening. Coordinate the interface of eggs and vanilla, avoiding an overrun scenario to the best of your skills and abilities.

At this point in time, leverage flour, baking soda and salt into a bowl and aggregate. Equalize with prior mixture and develop intense and continuous liaison among inputs until well–coordinated. Associate key chocolate and nut subsystems and execute stirring operations.

Within this time frame, take action to prepare the heating environment for throughput by manually setting the oven baking unit by hand to a temperature of 375 degrees Fahrenheit (190 degrees Celsius). Drop mixture in an ongoing fashion from a teaspoon implement onto an ungreased cookie sheet at intervals sufficient enough apart to permit total and permanent separation of throughputs to the maximum extent practicable under operating conditions.

Position cookie sheet in a bake situation and surveil for 8 to 10 minutes until cooking action terminates. Initiate coordination of outputs within the cooling rack function. Containerize, wrap in red tape, and disseminate to authorized staff personnel on a timely and expeditious basis.

Output:

Six dozen official government chocolate chip cookie units.

The cartoon on page 13 puts it another way.

Not to leave the legal profession out, an equally playful example of legalese follows, this one from *The Legal Guide to Mother Goose:* [4]
Mother Goose told it like this:

> *Jack and Jill went up a hill*
> *To fetch a pail of water*
> *Jack fell down and broke his crown*
> *and Jill came tumbling after.*

In legalese, it would go like this:

> *"JACK AND JILL"*
> Accident Report
> *The party of the first part hereinafter*
> *known as Jack . . . and . . .*

[4] *The Legal Guide to Mother Goose,* trans. by Don Sandberg (Los Angeles: Price/Stern/Sloan 1980), pp. 6-11. Reprinted by permission of Price/Stern/Sloan Publishers, Inc. © 1980. All rights reserved.

The party of the second part hereinafter known as Jill . . .
Ascended or caused to be ascended an elevation of undetermined height and degree of slope, hereinafter referred to as "hill."

Whose purpose it was to obtain, attain, procure, secure, or otherwise, gain acquisition to, by any and/or all means available to them, a receptacle or container, hereinafter known as "pail," suitable for the transport of a liquid whose chemical properties shall be limited to hydrogen and oxygen, the proportions of which shall not be less than or exceed two parts for the first mentioned element and one part for the latter. Such combination will hereinafter be called "water." On the occasion stated above, it has been established beyond reasonable doubt that Jack did plunge, tumble, topple, or otherwise be caused to lose his footing in a manner that caused his body to be thrust into a downward direction. As a direct result of these combined circumstances, Jack suffered fractures and contusions of his cranial regions. Jill, whether due to Jack's misfortune or not, was known to also tumble in similar fashion after Jack. (Whether the term, "after," shall be interpreted in a spatial or time passage sense, has not been determined.)

There is a movement afoot to eliminate legalese, gobbledygook, and the like. And it's gaining ground, thanks to the government. It's called the Plain Language Movement, and many states have passed laws which decree that legal documents, like insurance and banking policies, must be written so that the customer can understand (without the aid of legal counsel) what he or she is signing. A good example of the changes brought about by the Plain Language Movement follows:

THIS IS A PLAIN LANGUAGE POLICY.[5] PLEASE READ IT CAREFULLY. This policy is a legal contract between you and us. We suggest you read the policy summary first to get a general idea of the coverage. Then check the Policy Information Page. It shows the type and amount of coverage you have purchased. Be sure the information there is correct.

If you have any questions or wish to make any changes, talk to your Agent or get in touch with us.

YOUR 10 DAY RIGHT TO CANCEL: You can cancel this policy for any reason within 10 days after you receive it. Just mail it or deliver it to our Agent or Home Office. We will refund any premium you have paid.

THIS IS AN ANNUAL RENEWABLE AND CONVERTIBLE LEVEL TERM INSURANCE POLICY.
The proceeds are payable when the insured dies while the policy is in force. Premiums are payable during the insured's lifetime as stated on the Policy Information Page. We may charge a premium which is lower than the Guaranteed Maximum Premiums shown. This Policy contains the right of renewal to age 99 and the right to convert coverage prior to the policy anniversary nearest the insured's 65th birthdate. This Policy may be exchanged for a new policy every four years subject to policy provisions. There are No Dividends on this Policy.

By the way, the insurance company was kind enough to send the following example on page 16 illustrating the way their policies used to read. Aren't you glad they changed?

The Plain Language Policy points the way for others to follow. Unfortunately, the bad still far outweighs the good. If you're in doubt, skip to the chapter entitled "Cases in Point."

As for your own writing, it's probably true that you can't always avoid complexity, especially when you're writing on a technical subject. But when you have a choice, opt for simplicity, both in word use and sentence structure, so that when your choices are limited, you won't already have used up options and be forced into needlessly complex constructions. (See pp. 61–62, "Choose Simple Words.")

---

[5] American Health and Life Insurance Company, Baltimore, MD.

POLICY NUMBER

ANNUAL PREMIUM

$ 28.65

NATIONAL
OLD LINE LIFE INSURANCE
COMPANY

Amount $ 1000.00

HEREBY AGREES TO PAY

Age 19

- - - - - - - - - - - - -ONE THOUSAND AND NO/100- - - - - - - - - -Dollars
(the Face Amount of This Policy)

at the Home Office of the Company, Wichita, Kansas

immediately upon receipt of due proofs of the death of

MILLARD L. BILLING _____ the Insured,

ALBERT BILLING--FATHER & EDITH BILLING--MOTHER, SHARE AND SHARE ALIKE,
TO IF LIVING, OTHERWISE TO THE SURVIVOR

of the Insured (with the right on the part of the Insured to change the beneficiary as herein provided) if living, otherwise to the executors, administrators or assigns of the Insured, this policy being then in full force.

CONSIDERATION

This policy is issued in consideration of the application therefor, copy of which is hereto attached and made a part hereof and of the payment on or before the delivery hereof of

- - - - - - - - - - - - - -TWENTY-EIGHT AND 65/100- - - - - - - -Dollars

as the premium for one year's term insurance and the advance reserve required by law, if any,

beginning on the SEVENTH _____ day of SEPTEMBER _____, 19 33, which is the date of this policy, and in further consideration of a premium of like amount payable on or

before the SEVENTH _____ day of SEPTEMBER _____ in every year thereafter until premiums have been paid for twenty full years, including the first, or until prior death of the Insured.

The benefits, conditions and privileges stated on the following pages hereof are hereby made a part of this contract as fully as though recited at length over the signatures hereto affixed.

In Witness Whereof, The National Old Line Life Insurance Company has caused this policy to be signed by its President and Secretary at its Home Office in the

city of Wichita, Kansas, this SEVENTH _____ day of

SEPTEMBER _____, 19 33

SPECIMEN COPY

Attest:

Secretary.

President.

OFFICE OF
SUPERINTENDENT OF INSURANCE
TOPEKA, KANSAS

This policy is registered and secured by a pledge of bonds or notes and mortgages on real estate deposited with the State Treasurer of the State of Kansas in an amount equal to the full legal reserve on this policy.

Topeka, Kansas, _____ SEP 1933 , 19 ____

By _____
Assistant Superintendent.

Superintendent.

DIB   Form No. C-1-28.   Twenty Payment Life with 5 Coupons, Participating—Five Year Distribution.

Neologisms

Frequently, using legalese, bureaucratese, and so forth, leads to the creation of new words, or neologisms. So what's new about neologisms? Nothing, really. They've been around forever and serve the useful purpose of providing new words to describe things, especially new things. Many recent additions to our language come from the space program, the computer field, and other advanced technological areas. Many others come from more ordinary activities. Two recent additions now in most dictionaries are *debriefed*, used to describe the follow-up sessions for a returning dignitary or military person and *deplaned*, used to describe the activity of leaving a plane.

Sometimes, however, neologisms are not created out of necessity or for convenience, but as a subterfuge, a way of seeming to say something without actually saying it. When there was talk of raising taxes, a most unpopular issue, a White House spokesperson called it *revenue enhancement.* Since *revenue enhancement* sounds like a good thing, the term can be used without invoking all the fuss that the term *tax hike* might create. In another burst of creative vocabulary management, someone decided to change the name of the White House *Crisis Management Team* to *The Special Situation Group* to eliminate the negative implications of the former name. In recognition of this trend, The National Council of Teachers of English gives what it calls an annual Doublespeak award. A recent winner was the State Department for issuing a directive to remove the word *killing* from its official reports on the status of human rights worldwide and to replace it with "unlawful or arbitrary deprivation of life." The Pentagon was also cited for calling *peace* "permanent prehostility" and *combat* "violence processing."

At other times, new words or expressions are coined to replace a word or expression that no longer conveys an accurate picture of a situation. For instance, when the friendly skies began to fill with stewards as well as stewardesses, a new term, *flight attendant*, was developed to eliminate sex discrimination. The same rule applies to the name change from *mailman* to *postal worker* and *mail carrier*. In other instances, a name change results from the desire to increase the stature of a job by giving it a more prestigious name. For instance, garbage collectors have attempted to raise their status by changing their name to *sanitation workers* and often, now, janitors are known as *sanitation engineers.* Also, when gas stations did more than pump gas, they became *service stations*. Now, because of economic pressures, many have turned into *self-service stations*. The other day I saw a "help-wanted" sign in front of one of these service stations, which read, "Driveway salesman wanted"!

While lots of neologisms come from the government and big business, still others come from just plain folks. The worst terms today are the *-ize* words. You know (and possibly love) some of them like *prioritize, finalize,* *definitize*. You probably don't think anything about your use of these sorts of words, yet you won't find most of them in any dictionary. Although some may eventually receive official sanction in the dictionary, most are not considered proper English.

About the *-ize* syndrome *The American Heritage Dictionary* states, "The practice of turning nouns or adjectives into verbs by adding *-ize* is an ancient and useful one. It has created many standard words, such as *Americanize, criticize, formalize, nationalize, specialize,* though some of these were met with resistance when they were first introduced. But this practice has also given us such words as *finalize* and *concretize,* which still bother many people, and a great many linguistic experiments of questionable value, such as *envisionize* and *reprivatize.* New coinages of this sort should be used with great caution until they have passed the tests of utility, permanence, and acceptance by good writers."[6]

Words that have not "passed the tests of utility, permanence, and acceptance by good writers" but that have appeared in reports include the following:

formulize	clericalize
folderize	mezzanize
initialize	facilitize
unitize	incentivize
customerize	synopsize
modularize	operationalize
robotize	conveyorize
ruggedize	style-manualize

First cousin of the *-ize* syndrome is the *-tion* syndrome. The ending *-tion* works the same, by attachment to words to form longer words.

Using some of the *-ize* words above, we can form

folderization	customerization
initialization	modularization
digitization	robotization
unitization	operationalization

Others I have seen in reports include:

facilitization
prioritization
maximization
optimization
segmentization
conscientization

---

[6] *American Heritage Dictionary,* 2nd college ed. (Boston: Houghton Mifflin, 1982) p. 682. Copyright © 1982 Houghton Mifflin Company. Reprinted by permission from the *American Heritage Dictionary,* Second College Edition.

I have even seen *suboptimization* and *deproliferation*!

Next of kin is the *-ity* family as in *operability, performability, promotability, manufacturability,* and so forth. The following sentence from a technical report points up a typical example of this construction:

> *Superimposed on this discussion of satisfying the system requirements are the considerations of cost, performance, interface manageability, adaptability, and system maintainability and reliability.*

How did we develop this syndrome? This is how. A writer wants a verb for a construction like *to set priorities.* A single word isn't handy, so the writer invents one, *prioritize.* Now the writer has a verb. Next, the writer needs a noun, so, not finding one handy, creates *prioritization.* But whatever happened to the perfectly good noun *priority?* It got lost in the shuffle. And what emerges to take its place is a needlessly complex substitute.

Jargon

Jargon—your group's special vocabulary—can be a problem in technical or business writing. What is everyday shoptalk to you and needs no explanation is practically a foreign language to an outsider. Since most writing is done for outsiders (those not intimately connected with your work), most of your readers won't know your daily jargon. Witness the computer field as a good "bad" example of an industry with specialized jargon. For people familiar with computers and their special terminology, words like *input* and *output modes, hardware* and *software, CRT's* and *VDT's, bits* and *bytes, floppy disks,* and *interface* pose no problems. But those unfamiliar with such terms would rather about-face than interface. If a writer uses such terms outside his or her immediate professional circle, these terms must be explained. If they can be avoided, they should.

Acronyms fall within this group, too. Useful if known, confusing if not known. Typical examples of well-known acronyms include RADAR and NASA. In fact, you may not even know what they stand for since they're so commonly used as acronymns. RADAR is RA(dio) D(etecting) A(and) R(anging) and NASA is N(ational) A(eronautics) and S(pace) A(dministration). But acronyms, too, can get out of hand and even sometimes get silly as, for example, the acronym POTS demonstrates. What does it stand for? Plain Old Telephone Service!

Mood and Voice

The English language has both mood and voice. These have nothing to do with whether a person is elated or depressed or how loudly one speaks. Mood has to do with the way in which a statement is expressed. English has three moods: indicative, imperative, and subjunctive.

- *Indicative* is for most statements and all questions.

  EXAMPLES: I need the report by Friday.

  Do you need the report by Friday?

- *Imperative* is for commands and requests (where the subject *you* is missing but understood)

  EXAMPLES: Please give me the report by Friday.

  Give me the report by Friday!

- *Subjunctive* is for conditions contrary to fact, for wishful thinking, or for requests following the word *that*. The subjunctive mood has very limited use, but it is most often misused.

  EXAMPLES: I wish I were company president. (*Not:* I wish I was company president.)

  If I were president, I'd make a lot of changes around here. (*Not:* If I was president, I'd make a lot of changes around here.)

  He requests that you be on time tomorrow. (*Not:* He requests that you are on time tomorrow.)

Voice is another story. Voice refers to the use of a transitive verb, that is, a verb that needs an object. If the subject of the sentence performs the action indicated by the verb and the verb transfers that action to the direct object, the voice is *active*. If the person or thing receiving the action becomes the subject so that the subject is now being acted upon, then the voice is *passive*.

Maybe an example will help. In the sentence "John throws the ball," the subject (*John*) acts *(throws),* and the action is transferred to the direct object (*ball*). That's active voice. To change the sentence to passive voice, the object must become the subject and be acted upon by the agent (in this case, *John*), which may or may not be included in the sentence. One could write "The ball is thrown" or "The ball is thrown by John." In the first sentence, we don't know the agent of the action. In the second sentence, it takes two more words to provide the agent, and we have to wait until the end of the sentence to find out who it is. That's the problem with passive voice: it takes more words to say the same thing, or sometimes to say less, and the sentence lacks authority since the agent performing the action is not the subject of the sentence. If one chooses to eliminate the prepositional phrase *by John*, then the agent of the action remains unknown.

Now you probably see why the passive voice is so popular. Take the following sentence: "It was decided to fire John Jones." Passive voice is handy because no one is responsible for the action. It's also very much overworked and misused, often without the writer's being aware of it. But it's easy to spot since the passive voice construction contains any form of the *to be* verb and the past participle, the *-ed,* or *-en* form of the verb. Examples: *was received, will be notified, is written, were instructed.* You'll want to be on the lookout for passive voice constructions since they lead to weak, wordy writing that often obscures meaning.

Granted, there are good, legitimate uses of the passive voice. For example "The building was damaged by high winds." In this example the agent of destruction is not as important as the thing damaged since the wind didn't act with motivation. The important fact is that the building received damage. If, however, the insurance policy on the building covers wind damage but not water damage, the agent is important, and the sentence should read, "High winds damaged the building." If the agent doing the acting is responsible for the action, it should hold the subject position in the sentence and that means active voice.

Personal Pronouns

Once it was considered improper to interject one's person into professional writing. It was thought unprofessional. Journalists, for instance, would write, "This reporter recalls knowing . . . " when describing a personal acquaintance. The same held true for scientific and technical writing. The use of pronouns like *I, we,* and *you* was thought to reduce the objectivity of the writing. Thus, if the writer wanted to describe an experiment he or she had performed, the writer was forced to use the passive voice: "An experiment was conducted" or the more awkward "This writer conducted an experiment."

Such practice has gradually changed. It is now considered appropriate to use personal pronouns when needed. The key is necessity. It is not necessary, for example, to write, "I think the group did an excellent job." If you think so, then the sentence "The group did an excellent job" says so. The same holds true for "I feel" constructions. They should be eliminated in such sentences as "I feel that we should all work late tonight." The meaning is conveyed with fewer words in the sentence, "We should all work late tonight."

If, however, you have conducted an experiment, a survey, an interview, or the like and are reporting the findings, you can write, "I interviewed," "I determined," "I tested," etc. If more than one person did the work, you write, "We interviewed," etc. If the report addresses a specific reader or readers, you should use *you* or commands. In a memo requested by your boss, you might write, "As you requested, I began work on the project . . . ." For assembly instructions you might write, "First you pick up the L-shaped part and then you insert it into the L-shaped opening." Better yet, drop *you* altogether and use commands: "First pick up the L-shaped part and insert it into the L-shaped opening." Here, the subject *you* is understood.

Analogy

A useful device for conveying information is analogy: word picture. More specifically, analogy is the comparison of unlike objects through some point of similarity. Using analogies helps readers understand something unfamiliar by giving them something familiar to picture.

For example, if you tell your readers that a particular object looks like a safety pin or works like a corkscrew or is strung like a bow, they get the picture. The following example from a technical report shows a typical use of analogy: "The die works like an immobile, inverted cookie cutter, with the material dispensed over the die and the pressure applied by rubber rollers to stamp the ply." No matter how sophisticated the audience, analogy can help in getting across unfamiliar concepts.

National Geographic provides a beautiful illustration of analogy in the following description of the movement of tiny fish in lamplight at night: "There, shrimplike amphipods and krill dance as furiously as insects under a streetlamp. Larval fish dart toward them, like ghostly salmon leaping up waterfalls of light."[7] Can't you just picture it? If you can, it's because of the analogies the writer uses.

---

[7] Kenneth Brower, "A Galaxy of Life Fills the Night," *National Geographic,* 160, No. 6 (December 1981) p. 838.

Company Style Manuals

Most large companies have style manuals which describe the particulars of usage and format for company reports. Some are clearer than others. If your company has a style manual, you will probably have no choice but to follow it (let's hope that it sets a good example).

Smaller companies do not generally publish style manuals, relying instead on "standard practice" for procedures. Standard practice refers to the way something has always been done. Typically, that means that when you have to write a report, you'll go to the files to get an example of a report someone else wrote. The person whose report you use did the same thing when he or she had to write the report, and the example can probably be traced through the annals of the company. If, by chance, the original example was a good one, by now, more than likely, it looks and sounds old-fashioned. More commonly, though, the original was a poor example of communication practices, which means that from The Year One, everyone has been copying someone else's bad style. So, instead of reaching for that old report, make a clean break with the past. Just think: *your* report might set the new standard!

3
Know Your Audience

Who are you? Why are you writing? And for whom are you writing? No, this isn't the voice of your conscience speaking, but these are the questions you must ask yourself in good conscience before you begin to write.

Picture Your Reader

Before you can write anything, whether it's a note to the mail carrier, a memo to your staff, or a report to the company president, you have to know who you're writing to (or, more properly, to whom you are writing). If you're writing to a specific individual, as in the case of the note to the mail carrier, it's easy to identify and write to that person. If you're writing to a homogeneous group like your staff, that's also easy since you know them and know how to talk to them. But if you're writing to the company president or the marketing team or a large audience within the company, the task becomes more difficult because (1) you can't be certain that only one person or homogeneous group will read the report; (2) you have to consider who else might want the information in your report and how to address these unknown people; (3) the larger the report, the more likely it is that it will have a large audience; and (4) even specific memos addressed to specific individuals can receive wider distribution than you had originally planned.

An example from personal experience shows what I mean. A memo I wrote to a committee at my college about a curriculum change I was proposing led to a debate by that committee but no decision. As a result, the committee sent my memo to another committee, which took up the matter. The second committee also made no decision but sent my memo to the president of the college and to all the department heads. At that point, one department head wrote a memo countering the stance I had taken and sent a copy to me, to all department heads, and to the two committees involved, asking that the appropriate group make a decision. My memo, which was written in the spring of one year to a specific committee, had now made the rounds of the campus through all levels of management, had served as the basis for another memo from a different person, and was now back at the original committee. The point is that one must consider *all* potential readers when writing a report, letter, or memo.

Easier said than done, however. How do you know who might read your writing? Well, you don't know for sure, but you can make some educated guesses based on the kind of information you want to convey in your report, letter, or memo. Who might be affected by it? Who needs the information for decision making? For budget considerations? For marketing? For forecasting? Whoever might need any or all of the information in any piece of writing you produce is a potential reader of that writing and has to be included in your planning of it. To write to your entire audience, you must first divide it into two groups: primary readers and secondary readers.

Primary Readers

Your primary consideration is to your primary reader. The primary reader is generally the person who asked for the report or memo or the person or persons you have clearly in mind when you decide to write. Whether an individual or a group, the primary reader is easy to identify and write to because this person or group is the known quantity, the audience you anticipate. And if composed of more than one reader, the audience is unified in needs and interests.

Writing would be much easier if we could count on an audience consisting of a known individual or group. In fact, however, only a small percentage of what is written goes solely to a homogeneous audience. An even smaller percentage of what's written goes to an audience that knows just what we do, since there's little need to write to someone describing something he or she already knows.

Secondary Readers

More often than not in writing, one addresses a mixed audience, that is, a number of readers with various needs and understanding of the subject.

This audience is composed of secondary readers, those more distantly removed from the content of the message but nonetheless interested in or affected by it. If, for instance, you're recommending a procedural change in the office, the boss may pass your recommendation report on to higher management, or to purchasing if it requires any expenses for equipment, or to personnel if it requires any additional staff. All of these potential readers need to understand the contents of the message so that additional memos or phone calls are not needed to clarify any ambiguity or misunderstanding. In addition, all of the readers may not need all of the report, and the longer the report, the less likely it is that they'll all find it necessary to read all of it.

Prepare to Meet Your Audience

How, then, do you address all of your audience without boring those who already know certain aspects of the subject and without confusing those who need the basics explained? You do it in several of the following ways:

1. By organizing the material properly

2. By using headings and subheadings as guideposts

3. By incorporating forecasting statements to announce new material

4. By summarizing your material at the beginning of your report to give the readers an overview of the points you will cover

5. By clearly stating your purpose at the outset

All of these points are addressed in the following sections.

Readers Come in All Shapes

Each group of readers has different needs and interests based on the job, the background, the experience, and the education of its members. Usually, audiences break into the following groups: (1) general readers, (2) technicians, (3) engineers and technologists, (4) specialists, (5) managers, and (6) users. One of these groups contains your peers, the people you identify with. One of these groups contains your primary reader; the secondary readers come from one or more of the other groups.

GENERAL READERS

General readers are interested in the information in your report because they have the curiosity or desire to be informed. They are far removed from you in technical expertise, so explanations must be simple, yet thorough. They probably require a good deal of background information, and if you use illustrations, they must be simple enough to be understood by the otherwise uninformed. But don't be fooled. While general readers aren't experts in your field, they aren't blockheads, either. They could be, and often are, highly educated, but not in your area of expertise. Therefore, you must be guided by this maxim: If you can explain it, they can understand it. That means the burden of communication is on you.

TECHNICIANS

Technicians are experts, to one degree or another, in their particular area. Their interests are mainly in making things work. They generally are concerned with the "how" of things. Sometimes they use the design of engineers to build models. Often they maintain equipment. To do these jobs, they need lots of clear illustrations and explanations—practical explanations, not highly abstract, theoretical ones.

ENGINEERS AND TECHNOLOGISTS

Engineers and technologists are generally concerned with the "why" of things. Like technicians, they want to know how a thing works, but unlike most technicians, they also want to know why it works that way. How is it similar to or different from others of its kind? What are its applications? Explanations and illustrations should be more advanced for this audience than for technicians.

SPECIALISTS

Specialists are concerned with the steps leading to your conclusions. They want to know about the research techniques, how your research differs from that of others, what experiments you have set up, and so forth. They are very much interested in the "what" of your procedures and research methods. Because of their training, they can understand complex formulas and diagrams, and they will examine these very carefully. Results must be substantiated to satisfy their natural curiosity and skepticism; data must be provided.

MANAGERS

Managers are interested in decision-making. They want to know how the information presented in your report will affect growth and development of the company. They want to know what your recommendation will cost, what changes will be made, and how such changes will affect productivity. They want to see how this information fits into the big picture. They are most interested in your conclusions and recommendations. Even if they once had the same technical expertise as you do, they probably have lost at least some of it because, as managers, they now have other concerns. In fact, many managers don't have your technical background and never did, since their forte is dealing with people, administering, coordinating, and decision-making. They expect clear explanations that address their concerns. Your report must include technical data to support your findings, but most managers won't read the technical sections. Instead, they'll pass the report on to the appropriate departments for review.

What managers are after is your assessment of your findings. They may finally disagree with your recommendations, but they'll be able to do so intelligently if you've presented your findings clearly. They'll respect you for your work even if they decide on an avenue of action which differs from your recommendation. It's their ability to make a decision based on your report that makes your report useful to them.

USERS

Users want to know how to operate machines or equipment. They may have some expertise or they may have none. A user could be the machine operator who knows how to operate other types of machinery but has never operated the machine you describe. Users could be computer operators who have no technical expertise but who have to learn the in's and out's of the equipment they'll be using on the job. In both cases, users are people unfamiliar with the device or mechanism they must learn to operate. People in this category require clear instructions addressed specifically to their needs. They are not interested in theory, although they could profit from a general description of the equipment's capabilities or method of operation. Typical users are parents putting together Johnny's or Mary's bicycle on Christmas eve in the middle of the night, mechanics learning how to build or maintain or operate a new machine, secretaries being trained on a word processor, or children reading the instructions for a game. All of us are users at some point. And all of us have suffered the frustration that has made commonplace the expression, "When all else fails, read the instructions!"

Since we all become members of different audiences, depending on our need for and use of the material being presented to us, the writer must analyze who we are in each instance to write specifically for us.

4
Let's
Get Organized

Once you know your audience, you're ready to make some decisions about the organization of whatever it is you're writing. First you must ask yourself three questions: (1) What do my readers already know? (2) What do they need to know? and (3) What do they want to know? The answers to these questions determine the content and organization of your writing.

Whatever your readers already know, you don't need to tell them again. What they want to know must be the basis for the report, memo, or letter you're writing, and what they need to know in order to understand what they want to know must also be included. For example, if they need background information to understand the nature of your study, then even though they haven't asked for it, you must give them that information so that your report will be clear. Or if they want a recommendation, they may need to know what you rejected to understand fully how you arrived at the conclusion that led to your recommendation.

In addition to the three questions listed above, a fourth question also plays a part in organizing your report. What is the audience's attitude toward the subject? If, for example, your recommendations for streamlining office procedures will cause a staff cutback and require the purchase of expensive equipment, the affected members of your audience could react negatively. For your report to be effective, you must not only explain your recommendations but you must also *convince* your audience of the rightness of your recommendations in terms of company goals and their interests.

Order the Parts

The order of the parts of your report should be based on your readers and the subject matter. When the readers are likely to be hostile toward the information in your report, order your material chronologically. That is, begin at the beginning with a statement of the problem, need for a solution, your approach to a solution, and so forth. Build your argument step by step. By the time you reach the conclusions and recommendations, the readers will understand the reasons behind your recommendations and will be better able to accept them. Chronological arrangement also works well for presenting highly technical information to a non-technical audience since it allows the readers to build understanding of the points made in your report gradually as they move toward the conclusion. For instance, if you were describing sheet metal fabrication equipment to an audience of accountants, you would need to explain each piece of equipment, its capabilities, and the advantages of purchasing each so that the audience would understand the importance and desirability of each purchase. Beginning with a statement such as "We need sixteen trillion dollars for equipment purchase" is likely to receive a negative response no matter how wonderful or necessary the equipment might be.

Conversely, if the information is likely to be readily understood by your readers, such that lengthy explanations would bore and frustrate them, then the chronological approach is not a good choice. Rather, you begin with recommendations or conclusions and work back through the facts leading to the recommendation, the background information, then the statement of the problem, etc. Since the most important information is up front, it has the best chance of being read, and the readers won't have to be dragged through background and explanations they might not need or want. This approach only works, however, if the audience can truly understand the "ending" before the beginning or the middle. An example of this approach might be a report describing a change in procedure that the audience can understand without explanation. For those who need more, the writer can elaborate on the methods used to develop the change, the options considered, the criteria for selection of the best solution, and the problem that led to the need for the change in the first place.

Finally, for "mixed" audiences, those representing diverse technical backgrounds and interests, it is best to start with an overall summary of the central findings of your report or the chief points you present (if no conclusions are drawn) and follow that with a chronological discussion. While any report is enhanced by a summary (more will be said about that later), the summary for a mixed audience must be particularly good since many people will read only this section. This method of organizing reports probably has the widest application, since most audiences are "mixed."

What Readers Read

What may come as a surprise, but shouldn't, is that very few readers will go through your whole report from start to finish. After all, it's not a novel they're reading, and even when your material is presented in a lively manner (writing need not be dull to be technically accurate), you shouldn't be offended if you lose readers along the way. This will happen in the best of reports precisely because they *are* well written. As readers get what they need, they drop out. Unfortunately, the worst reports have the same results, but in this case readers drop out from frustration, not satisfaction.

The key to good organization is knowing what readers need to read. A Westinghouse study of the reading habits of managers indicates that 100 percent of the managers read the summary of a report. After that, 80 percent read the introduction, 15 percent the body, and 50 percent the conclusions and recommendations. If there is an appendix, 10 percent read it.[8]

What do these figures mean? First of all, by extension to others besides managers, they mean that audiences only read what is relevant to them. Second, they mean that you must give your readers what they need and make it easy for them to find it. How do you accomplish this? With headings and subheadings to indicate divisions, with bulleted and numbered lists, with forecasting statements, and with a good summary to give the proper overview. These are described in the following sections.

Head 'em Up with Headings

Any report, no matter how brief, can be enhanced with headings. The longer the report, the greater the need for them. Headings do two essential things: (1) They tell the reader what information is contained under them, and (2) they break up the text into bite-sized chunks.

The first point seems fairly obvious but is quite often ignored by writers. It may be that writers don't think about their audience at all, or that they find writing headings a nuisance. Yet, we've all had to wade through reports with no road signs, and we've all been waylaid without getting what we really needed from the report. Headings point the way. If the headings are descriptive and frequent, they tell the reader whether to read a section or skip it. The reader is saved frustration. The writer gets the message across.

---

[8] Richard W. Dodge, "What to Report," *Westinghouse Engineer,* 22, No. 4-5 (1962) pp. 108-111.

In addition, headings break the heavy black line of type. They provide restful space. There's nothing more depressing to a reader than having to flip through a report of unremitting type that extends from margin to margin and from page to page. Nothing catches one's eye; nothing gives one pause or even suggests what is to come. If the reader has any choice, he or she will put the oppressive-looking report at the bottom of the stack of things to read, where it's likely to remain because there never is enough time to tackle such a thankless task. All the writer's effort, more often than not, goes unrewarded because the report remains unread.

Choices in formats for headings vary from company to company, textbook to textbook. If you do not have specific instructions from a textbook or company style manual, the following guidelines and the figure on page 35 will be useful. They describe one of many possible ways of ordering headings. Where specific instructions don't exist, any logical method that suits your needs is acceptable. The key to the use of any method is consistency.

FIRST-ORDER HEADINGS OR MAIN HEADINGS

First-order headings are used in formal reports to indicate the major sections. They are centered on the line, typed in capital letters, and may or may not be underlined. If the report is double-spaced, the headings are given a bit more white space by leaving three or four blank lines before and after them.

SECOND-ORDER HEADINGS OR SIDE HEADINGS

As the name implies, second-order headings or side headings appear in a secondary position to the main headings at the side on the left margin. They may be typed in capital letters or in upper and lower case. They may be underlined or not, with two or more blank lines before and after them. Second-order headings serve as main headings in short reports.

THIRD-ORDER HEADINGS OR RUN-IN HEADINGS

Third-order headings or run-in headings come at the beginning of a paragraph, so if you indent for paragraphs, you indent for these headings. They are underlined, typed in upper and lower case, and can be followed by a colon or a period. If the heading is a descriptive phrase, I use a colon after it. If the heading is a sentence, I use a period after it. In either case, the text follows on the same line as the heading.

Most reports don't require further subdivisions, but if they are needed, you can use bulleted or numbered lists (described on p.36). Another possibility is to make the main heading a spaced, centered heading like:

FIRST-ORDER HEADING

— or main heading

— or side heading

Second-Order Heading

— or run-in heading

Third-order heading

M A I N   O R D E R   H E A D I N G

Then, the next heading would be centered without spaces between the letters.

M A I N   O R D E R   H E A D I N G

Using this method, you can work with four levels of headings.

The Numbers Game

Some writers prefer the double numeration or ANSI (American National Standards Institute) system for headings. Many government agencies require this system. Main headings are numbered 1. 2. 3. etc. Subheadings are numbered 1.1 1.2 1.3 etc., with infinite subdivisions permissible, along the same lines: for example, 1.1.1.1.1.

In the ANSI system, section numbers are still followed by descriptive headings, but the numbers make it easy to organize very large reports, to make cross-reference to other sections, to match illustrations to numbered sections, to help identify chapters, and so forth. Also, since the numbers change with each new section, you don't have to re-number the entire report if you want to add pages in certain sections.

Some writers prefer an outline system, placing Roman numerals (I, II, III) before main headings and capital letters and Arabic numbers (A, B, C 1, 2, 3,) before subheadings. Others put numbers before main divisions and number the paragraphs successively within each division. For example, 2.65 would indicate the sixty-fifth paragraph in Section 2.

The point is, whichever system you choose, you should use consistently. Your efforts will be rewarded by increased communication.

Bullets—Shoot Straight

Writing should shoot straight. One of the techniques for straightshooting is to use bullets. Bullets organize ideas graphically into lists, provide soothing white space, and suggest the correlation between points.

Take the following paragraph in standard prose:

To build this model you will need three pieces of paper, a bottle of glue, a ruler, a pair of scissors, and a compass.

Look how much easier it is when you use bullets.

To build this model you will need the following:

- three pieces of paper
- a bottle of glue
- a ruler
- a pair of scissors
- a compass

Granted, the example is simple; nonetheless, the version with bullets is far easier to understand than the one without them. The more complex the material, the more useful the bullets become.

Numbered Lists

Numbered lists do much the same thing as bulleted lists except that numbers allow you to rank the information. You can indicate the ranking by telling the readers that the list is from large to small, most important to least important, or first to last. Numbers give each item in the list a position in reference to the others. When this is important in presenting your data, use numbered lists. The following example illustrates:

Before:

If we buy the proposed system, we will have to provide Smith with a schedule of all reports needed, tell Smith how to enter the data, give Smith a flow chart to show computations we need, submit all our data with codes on typed forms, and provide the data at least two weeks before the due date for the report.

After:

If we buy the proposed system, we will have to do five things:

1. provide Smith with a schedule of all reports needed
2. tell Smith how to enter our data
3. give Smith a flow chart to show computations we need
4. submit all our data with codes on typed forms
5. provide the data at least two weeks before the due date for the report

Numbers can also be used in standard sentence construction.

Example:

I recommend that you (1) turn down the Smith proposal, (2) notify Smith that we do not accept his offer, and (3) authorize me to work up a plan for in-house operation.

The Style of Lists

A word (or two) about the style or mechanics of lists is probably in order here. How do you punctuate a list? When do you capitalize? How do you begin a list? There are no hard and fast answers to these questions beyond the golden rule of consistency. But here are some pointers.

In the following example—

If we buy the proposed system, we will have to . . .

—many people would put a colon after the word *to*. I don't recommend a colon here because the items in the list complete the thought, and if I had written the sentence without the numbers, I wouldn't use a colon. Still, the practice is widespread. (See the discussion of the colon on pp. 103–104.)

But in the following example—

If we buy the proposed system, we will have to do six things:

—I would, as the example shows, use the colon at the end of the introductory line because the colon correctly follows a complete thought. In either case, the first item in the list and every item following it should be in the grammatical construction required to complete the sentence.

In both examples shown above the reader anticipates a verb, so each item on the list will have to begin with a verb. As the list on page 37 shows, each does.

However, if I had changed the lead-in to the list to read—

In buying the proposed system we will need
- schedules of our reports
- information on data entry
- a flow chart of computations
  etc.

—the sentence would require nouns to complete the thought, so each item on the list must begin with a noun to make all items complete the thought in the same way.

Whether you use bulleted or numbered lists or run your list into the sentence, you must use parallel construction for each item on the list. Parallel construction means that the information you present in a list must be in sentences or phrases that are alike grammatically. Parallel construction allows the reader to absorb related data easily, since the information is presented in the same grammatical form. The reader doesn't have to sort information to understand the correlation between points because you've presented the information in matching constructions. In the example above, you can see parallel structure in the use of nouns to begin each item in the list: *schedules, information, flowchart.* Also, the construction is consistent: all phrases— not a mixture of phrases and sentences. (For more information on parallel structure, see pp. 64–65.)

Another consideration with lists is how to begin each item on the list— with a capital or lower case letter? Usage varies, but some style manuals suggest that if the list is composed of full sentences, each item should begin with a capital letter. If the list contains phrases, each item should begin with a lower case letter. Since the examples above contain phrases, I use lower case letters to begin each item on the list. If I had used capital letters, it wouldn't be wrong. It just wouldn't be as logical since the items aren't complete sentences. Some style manuals indicate that all lists should begin with a capital letter. This certainly makes for consistency in the lists and eliminates the problem of making a decision. The choice is yours.

Next, there is the question of punctuation in lists. Should you end each item with a comma, semicolon, or period . . . or nothing? Again, usage varies. If the list consists of complete sentences, you could end each sentence with a period, or you could choose to leave the punctuation off.

> *Example:*
>
> In buying the proposed system we have the following obligations:
>
> 1. We will need to provide Smith with a schedule of all reports[.]
>
> 2. We will need to tell Smith how to enter our data[.]
>
> 3. We will need to give Smith a flow chart to show compuations[.]
>
> 4. We will need to provide the data at least two weeks in advance[.]

The [.] indicates that the period at the end of each item is optional.

If the list is a series of descriptive phrases, you have several choices. You can use commas after short descriptive phrases or semicolons after long descriptive phrases that contain commas within them. If you use commas or semicolons within the list, you place a period after the last item in the list.

Example:

If we buy the proposed system, we will have to

1. provide Smith a schedule of all reports needed[,]

2. tell Smith how to enter the data [,]

3. give Smith a flow chart to show computations we need [,]

4. provide the data at least two weeks before the due date [.]

The [ ] indicates that the punctuation is optional. If you put nothing at the end of each item in the list, you also finish with nothing.

Finally, if you run your numbered items into standard sentence construction, you punctuate between the numbered items the same way you would if the numbers weren't there. So, in the sample sentence below, I use commas to separate the items in the list, which is what I would use if the numbers weren't there.

I recommend that you (1) turn down the Smith proposal, (2) notify Smith that we do not accept his offer, and (3) authorize me to work up a plan for in-house operation.

What's The Forecast?

You'll run into foul weather if you don't tell your readers what to expect. A forecast describes the future. In technical and business writing, it tells the readers what material is up ahead. Thus, readers can decide whether they need to read the information or move to another section.

I use forecasting statements throughout this book to let you know what I am covering. Each one summarizes the material in advance, gives you a picture of the whole, and indicates the divisions. Have you noticed them? Maybe not. They shouldn't jump out at you. But they should help you anticipate the arrangement of new material. And since I can assume that you're still reading, it's possible to assume that the forecasting statements helped!

The following is a forecasting statement for the upcoming sections of this book: Having looked at the ways you can divide information for your audience and keep the readers reading only what they need and want, let's now look at the most important part of any report: the summary. (Note that I used that forecasting statement not only to look ahead but also to look back. In this way, it provides transition between divisions, reminding you of where you've been as well as indicating where you're going.)

The Summary— The Heart of the Matter

Suppose a report began with a statement like this:

"The main thing to remember in making a decision is that we have the cooperation of all parties involved." You'd say, "What is the writer talking about?" And you'd be right to be confused. But you'd be surprised at how many writers begin reports—if not always so blatantly in the middle of things—with the assumption that the person requesting the report knows everything that has come before the writing of the report.

Such an assumption errs in several ways: (1) it assumes that the reader is going to recognize immediately the nature of the report when it should cross his or her desk; (2) it assumes that this person will recall all the pertinent points leading up to the writing of the report; (3) it assumes that only one person or group will be reading the report.

An error of the opposite kind is committed as often. In this case, the writer begins at the beginning, the very beginning, and drags the reader along, finally presenting the important and long-awaited conclusions at the end of the report. Since this approach is useful only for presenting information which will not be understood or readily accepted by totally uninformed or biased readers and since most reports are not written for either of these categories of readers, this approach generally misfires. Often the problem is that the writer assumes that readers want to know everything. As we've discussed earlier, this is generally an incorrect assumption. Readers only *want* to know what they *need* to know and no more.

Both kinds of errors can be corrected with the addition of an introductory summary. The summary has several purposes:

1. It orients the reader to the nature of the report.

2. It states the writer's purpose.

3. It states the key points covered in the report.

4. It spells out any conclusions and recommendations reached.

A summary is not to be confused with an introduction or an abstract. Each has a separate function. An introduction, which generally follows a summary, contains some of the same information as the summary—statement of subject and purpose, scope of the problem, and points covered in the report— but it does not generally indicate the outcome or conclusions reached. Rather, it describes the approach the writer takes in solving the problem posed and the arrangement of the information in the report.

The abstract, which is often confused with the summary, is generally thought of as a prose table of contents. As such, it outlines the key points to be covered in the report but does not state the writer's position on these points. Many journal articles and formal reports contain abstracts to indicate to readers the kinds of topics covered in the article or report so that readers can determine if they need to read it. An example of an abstract for a formal report follows:

> Auto racing is an extremely popular spectator sport, but few fans are familiar with the work required to prepare a race car for the track. "Racing Wrenches: The Tasks of Mechanics at Car Races" describes for the average racing fan the duties and responsibilities of the race car mechanic. It traces the activities of the racing mechanic through the major stages of his work on the race car: first at the shop, then at the track before the race, and finally during the race. It concludes with recommendations about types of people who make the best race car mechanics.

When you read this abstract you know the topics the author will discuss, but you don't know the content, approach, or opinions of the author. You have to read the report to find these out.

Summaries differ from abstracts and introductions in giving the reader the whole story in a nutshell. A good summary should satisfy the readers. They *may* choose to read further, but they may not. If they do not, they should have the gist of the report from the summary alone.

In a one-page memo or short, informal report, the summary could be a sentence or two. In longer reports, it will be longer. A rule of thumb is that it should not exceed one typed page. Oftentimes, even in formal reports of 100 pages or more, a good summary might be only a paragraph or two.

A summary must include all the points mentioned above: the nature and purpose, key points, and conclusions and recommendations reached. It should not say, "This report includes conclusions and recommendations." That's the style of an abstract. Rather, it should say, "I recommend that we begin Phase II of the development project on June 1."

A summary *always* increases the usefulness and comprehension of a report, long or short. Without the summary, a report has no anchor, no focus. Yet it is often absent. Why? For one thing, it's hard to write. It requires an overview of the report itself when you know where you've gone and can tell others where they will be going. For another, it must be concise, yet it must include the essential points of the report. That requires some astute decision-making on the writer's part. Additionally, for those readers who will read beyond the summary, it must include information which is discussed in detail in the report, but it should avoid using the same words and phrases to keep the reader from experiencing a sense of *déjà vu,* or having been there before. This requires a certain imagination and a flexible vocabulary.

Despite all the difficulty of writing a good summary, what's hard for you as the writer will make it easy for your reader, and it's results you want. So spend the extra time to write the summary. It could be the most important part of your report.

Here are two examples of summaries for a short recommendation report:

> In my two years with the company, I have recognized the need for increased efficiency and economy in performing routine drafting and conceptual designs. I recommend that we upgrade the drafting department by purchasing a computer-aided drafting (CAD) machine. This report describes the type of electronic graphic system I recommend and the savings it can provide. The report does not recommend a specific brand at this early stage.

*   *   *   *   *

> When we began the job of selecting a word processor, it was immediately apparent that there was no possibility of examining all the systems available. Further, the technology is developing so rapidly that we cannot expect to choose a state-of-the-art system. Hence, we have confined our investigation almost entirely to systems with which the committee has had some experience. After examining five systems and narrowing the field to three, the committee recommends the purchase of the BRAND X with the BRAND Y printer.

The summary for a long report is no different, except that it is generally a bit longer than one for a short report. The following is a summary from a long informative report:

> Home buyers should consider installing solar relief heating systems in their new home for the following reasons:
>
> 1.   long-term savings
>
> 2.   higher resale value as conventional energy costs increase

3. energy conservation

4. ecological factors

> This report presents several types of active solar water and space heating systems available and describes the operation of each. The systems selected are designed to carry a certain percentage of the home's energy load requirements, thus conserving energy and saving money. This report shows home buyers how these systems can increase the value of their home without adversely affecting the environment.

The following summary from a long report recommends a solution to a problem:

> Observation shows that innovations in the cable television industry, specifically those connected with the 54-channel system, have caused numerous problems. Perhaps the most practical solution to these problems comes through careful and deliberate design.
>
> The cable television designer is charged with the responsibility of taking an array of information and compiling it into a neat, orderly, but practical working system, free from mechanical error. This task is becoming increasingly difficult because of the limitations of the new technology. Especially difficult is the design of the 54-channel system, primarily because of the dimension of two-way communication and the necessary limitation it adds.
>
> This report outlines the history of the problem which can only be resolved through the improvement of the present design practices and the establishment of new, more efficient practices.

How to Write a Summary

The first rule about writing a summary is to do it last. You know the rule about saving the best for last? Well, the summary should represent your best effort and you can write it best when you know what you want to summarize. It's too hard to know exactly how a report will come out before you write it, and if you write the summary first, it may not accurately represent the final contents of the report.

When you're ready to write the summary, try this technique:

1. State the problem in one simple sentence.

2. Write two or three more sentences to expand the explanation of the problem.

3. State your approach to a solution in a single sentence.

4. Write two or three sentences to expand the solution.

5. State the significant result.

6. Support this with a sentence or two.

7. Then edit *severely* to leave behind the kernel of your key points.

Why do all this writing and *then* edit? Why not just write it succinctly the first time? Because it is very difficult to produce a distillation of your thoughts in one draft. Instead, if you let the ideas flow, knowing you'll go back and separate the wheat from the chaff, you'll be able to write the summary more easily and effectively.

Here's a sample of the process in action:

1. There have been an increasing number of file folders lost or misplaced.

2. Papers that have been filed are either not found or are found only after someone realizes that a file is out without anyone knowing it. The search for file folders is time–consuming.

3. I recommend that we use "file–out cards."

4. This system requires that the person taking out a file folder write his or her name on a file–out card and insert the card in the folder's place.

5. This will reduce time searching for a file folder by indicating who has it.

6. It will reduce the number of missing file folders because the card will tell who has it.

7. *Edited version.* An increasing number of file folders have been lost or misplaced. To correct this situation, I recommend that we use "file–out cards." When a person takes a file folder out, he replaces it with a card bearing his name. This procedure will reduce the time spent searching for file folders and will reduce the number of missing file folders.

5
Graphics
Are Grabbers

Much could be said about graphics, but since the aim of this book is to be brief, I will leave the details to others who have written whole books on the subject. What follows are a few pointers.

Think Graphically

Not only should your words create clear pictures in your readers' minds, but wherever possible and whenever budgets allow, you should support prose statements with graphic illustrations.

In part, you're already doing this with headings, subheadings, bullets, and numbered lists, since these devices provide visual breaks in prose and point to the meaning of the text. Keep up the good work with tables and figures to support the data you present. Tables are illustrations that organize numbers and words into rows and columns. Figures are everything else: bar graphs, line graphs, pie graphs, organizational charts, maps, photos, etc.

A Picture Is Worth a Thousand Words

While a picture may be worth a thousand words, a table or figure needs explanation. Never assume that your graphics are self-explanatory, no matter how simply they are drawn or how obvious they appear to you. The text should always explain the relevant points of a visual aid, telling the reader the significant correlations to be observed. For example, if you're making a recommendation on a black and white TV for the break room of a factory and you've compared several brands in terms of cost, warranty on parts and picture tube, and other factors, your readers will be able to see comparisons more easily if you give them a table of the data. It might look like the one below. While it is clear to you how the items compare, you should still tell the reader what is significant in the table. You might write, "As Table 1 indicates, Brand D is the best buy for the break room because it has the lowest cost and best warranty."

Table 1: Television Set Comparisons

Brand	Initial Cost	Warranty Picture Tube	Warranty Parts	Warranty Labor	Special Features
A	$ 160	1 Year	1 Year	90 Day	Earphone Jack
B	$ 180	1 Year	1 Year	90 Day	Sleep Switch
C	$ 165	1 Year	90 Day	90 Day	—
D	$ 150	1 Year	1 Year	1 Year	Earphone Jack

Placement Is Important

Always place the visual aid immediately *after* the explanation. If there is room for it on the same page, it goes just after the paragraph in which it is discussed or within the same paragraph. It should not go before your discussion, however, because the reader won't know what it's doing there until you explain it. If it is too large to fit on the same page, it goes on the next page. If it is not shaped for vertical presentation, it is placed so that the reader need make only one clockwise turn of the page to read the visual aid.

Making one turn of the page won't bother most readers (although you should strive for vertical placement whenever possible). But hunting for the visual aid will bother them. Most likely, they won't bother. If you took the time to do a visual aid, put it where it will be read or examined. And if it's worth including in the text, it's worth discussing. If not, it belongs in an appendix, and you merely refer to it in the text.

Label and Number

Every visual aid should have a number to identify it. The only exceptions are in informal reports or reports which only include one aid. In addition to a number, each aid needs a label or caption which describes it. The number and label make for easy identification and reference. When a report contains both figures and tables, they are numbered separately (Figure 1,2,3, Table 1,2,3). In formal reports which contain a List of Illustrations, the figures are listed together and the tables are listed together. They can appear on the same List of Illustrations page, but they're separated by a subheading for figures and a subheading for tables.

Show and Tell

Tell the readers at the earliest possible moment that you've provided a visual aid for their benefit. Then discuss it. Then present it. It's very frustrating to the reader to wade through a complicated discussion, only to turn the page and find a visual aid which could have helped the reader's understanding if he or she had known it was there while reading, not afterward. Methods for introducing the visual aid include, "As Figure 6 shows" or "(see Figure 6)" or something similar.

Since anything you can visualize can be made into a visual aid, visual aids come in many forms. The most common types include tables, bar graphs, line graphs, and pie graphs. Choice is dependent on what you want to show.

TABLES

Tables work best for comparisons using words and numbers. The table on page 48 is a typical example. Tables have vertical columns and horizontal rows. Informal tables appear, as you might guess, in informal reports, and the rules governing them are—informal. They don't need a number or caption, they don't need a ruled frame around them to separate them from the text, and they have minimal internal lines separating columns and rows. White space will suffice to set them off. The virtue of informal tables is that they not only provide white space within, but they also provide white space all around to break up the text and give the reader a picture to look at.

Formal tables are, as their name implies, more formal than informal tables. Therefore, they have more rules governing their use. The main parts of a formal table are shown in Figure 1. They require a number and a caption, generally placed above the table, either at the left margin or centered. All columns and rows need headings. If the items in the headings are units of measure, these units must be indicated in the headings. Lines dividing the columns and rows are optional, with the trend being to leave them out for a less cluttered look. Footnotes at the bottom of the table are used to explain or clarify terms and to give sources. Footnotes are generally indicated by lower case superscript letters (a,b,c) or by an asterisk (*) if there is only one. Line headings are read across. Column headings are read down. Subheadings classify line headings and are also read down.

Figure 1: The Layout of a Table

Table x: Caption for the Table

Stub Heading	Column Heading[a]	Column Heading[b] Subheading Subheading	Boxhead
Line Head subhead subhead subhead Line Head		Field for units of data	Field
[a] footnote [b] footnote			

When a table continues to another page, the table number and headings are repeated on the second page for clarity. Often, the word *continued* appears at the bottom of the first page and at the top of the second page.

GRAPHS

Where tables show the comparison of distinct data, graphs report the relationship between things. Some show the relationship between things at a given time, others the changes over a period of time, and still others the relationship of parts to the whole. The most common graphs are line graphs, bar graphs, and pie graphs.

Like tables, graphs need a title and number. Since all graphs are figures, they are numbered consecutively as Figure 1, 2, 3, etc. Line and bar graphs have a vertical scale called the *ordinate* for the dependent variable and a horizontal scale called the *abscissa* for the independent variable. The dependent variable is the measurement that changes in relation to changes in the independent variable. The most common independent variable is time, against which the dependent variable changes as time progresses. So, as the years along the horizontal scale increase, the amount of something on the vertical scale—say, bushels of wheat or the number of cars produced in the United States—goes up or down or remains the same. The point at which the two scales come together is generally the zero point.

The scale for the vertical and horizontal axes should be incremental, that is, evenly divided, to present an accurate picture. It would be misleading, for instance, to plot the years 1950, 1955, 1960, 1965, 1966, 1967, 1968 along the horizontal scale because the reader generally pays closer attention to the picture than to the words and would therefore be likely to miss the shift in the scale.

Line Graphs are the most frequently used type in technical and business writing. They are best for showing the continuous flow of data over time. The relationship between points is plotted at the intersection of the abscissa and ordinate. When all the points are plotted, a line connecting them shows the relationship of all points on the graph. Figure 2 shows a model of a line graph. If accuracy is not as important as the presentation of a general picture, then the background grid can be left out. Sometimes, because of space restrictions, the graph doesn't start with zero, especially when the numbers plotted are very high. In this case, the lines between the vertical and horizontal scales are left unconnected to indicate what is called a "suppressed zero." When presenting the full line graph—including the plotted line, the frame lines around the graph and the grid—the ratio of widths of the lines you draw is 4:2:1 for plot: frame: grid.

Line graphs may also include more than one line to show the relationship between two sets of statistics. In this case, each line must be identified

Figure 2: Model of Line Graph

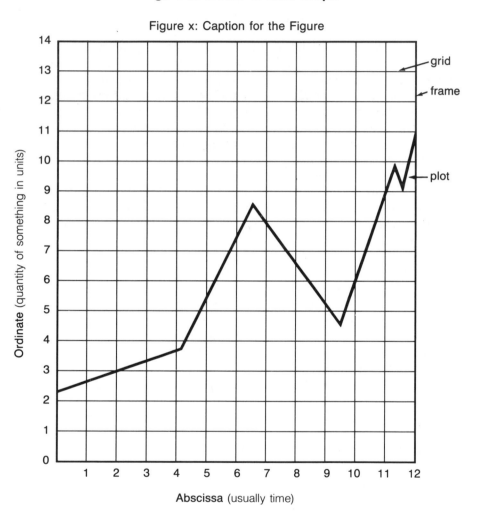

Figure x: Caption for the Figure

with a label or legend. Shading in or coloring the space between the two lines adds emphasis. The label describing the horizontal axis is placed below that axis. The label describing the vertical axis is placed at the top of that axis or along the side if the label is long.

Bar Graphs work best for charting discrete, or separate, data. They can be used to plot data for the same period or for different periods. For instance, a bar graph could show auto sales for one car maker for several years, or for several car makers for a given year, or for several car makers for several years.

Bar graphs can be vertical or horizontal. See Figure 3 for a model of each type. The choice usually depends on the data being depicted. Distance, for instance, should be plotted using horizontal bars; heights should be depicted using vertical bars. Sometimes, the orientation of a bar graph depends on the page layout. At other times, it depends on personal preference. Whether the bar graph is vertical or horizontal, it contains the abscissa and ordinate (discussed under line graphs). One additional rule is important to remember: the space between the bars should be smaller than the bars themselves to make it easy to distinguish the bars from the spaces. Shading or coloring the bars helps, too. As with the line graphs, bar graphs can have multiple bars joined together to show comparisons of several kinds of data in relation to a particular independent variable like time. In these cases, shading the bars to distinguish between them becomes essential.

Pie Graphs do what their name implies: depict data in the form of a pie. The beauty of pie graphs is their ability to show portions of the whole. They are often found in business reports because they provide a handy way to visualize percentages of profits over a year, budget expenditures, and so forth. Figure 4 shows a model of a pie graph. Starting at 12 o'clock, the pieces of the pie are arranged from biggest to littlest, clockwise. If there is a "miscellaneous" piece it comes last, even if it is bigger than another piece. The label for each piece, along with the percentage of each piece, can be inside the particular piece or outside the pie with a line drawn to the piece. All labels should be horizontal. Portions of the pie can be shaded to show the relative size of each piece. However, the biggest piece is usually not shaded. Also, the pie shouldn't contain more than eight pieces so that the reader can easily digest the information (pun intended!). The total pie, of course, must add up to 100 percent.

USING VISUAL AIDS

There are lots of other types of visual aids, including the following:

- Pictograms—variations of the bar graph, using pictures rather than bars to represent data (also called pictographs)
- Flow charts—generally the presentation of a process or procedure, often using computer programming symbols

Figure 3: Model of Horizontal and Vertical Bar Graphs

Figure x: Horizontal Bar Graph

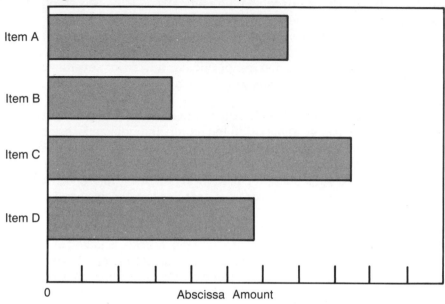

Abscissa Amount

Figure y: Vertical Bar Graph

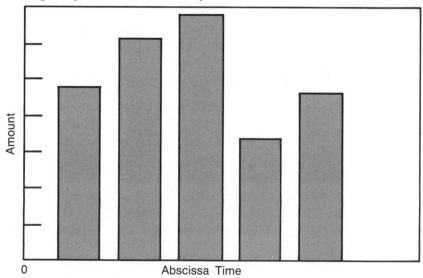

Abscissa Time

Figure 4: Model of Pie Graph

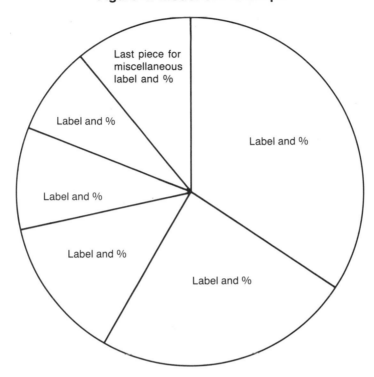

- Organizational charts—depictions of an overall arrangement or chain of command for the positions within an organization

- Schematic diagrams—abstract representations of mechanisms or systems that use lines and symbols for explanation

However, the visual aids described in this chapter—tables, line graphs, and pie graphs—appear most often in reports.

The rule of thumb with any visual aid is to match the picture to the audience. If the illustration is needlessly complex, general or managerial readers will ignore it. If it's absurdly simple, technical readers will ignore it. In both cases you've wasted your time (and probably your money, too). So match the visual aid to the readers' needs. Also, if you get the visual aid from another source, be sure and indicate the source at the bottom of the visual aid. It's not only polite and informative, but *legal.* If the report is to be distributed widely or published, you should also obtain permission.

6
Editing: The Garbage Collector

So, you've finished the draft of that report, and you're ready to have it typed. One more project off your desk and off your mind.

Not so fast. As tempting as it is to send the piece on for someone else to worry about, you still face one very important task: garbage collection, otherwise known as editing. Before you can give that report the final stamp of approval, you need to remove all the garbage. In this chapter I discuss the ways to recognize and remove weak verbs and nouns; strengthen weak beginnings; avoid redundancies, wordiness, and noun clusters; and generally make your report more readable and therefore more effective. While you should keep these editing points in mind when you're writing, you should not let them slow you down in the first draft. There's time for all that later—*after* you have something on paper to ponder.

Let the first draft flow as it will. Getting words on paper is a big hurdle. But don't bypass the next hurdle—editing—or your readers may stumble and fall in your garbage.

The Cooling–Off Period

"The sooner the better" is an adage that does *not* apply to editing. Why? Because the mind plays strange tricks on us, and one of them is to make us believe that what we *think* we've written is what we've *actually* written. Also, we think that all our sentences are clearly expressed because we knew what we *wanted* to say, and so, surely, we must have said it.

For these reasons, we can't edit or proofread properly because we can't see what's missing or misplaced. Also, we don't want to change anything because we're either too close to the writing, too tired of it (having just written it), or too pleased with it to see any flaws.

That's why the cooling–off period is so important. At the very least, it should be an hour; at best, it should be overnight or several days. Of course, most of us don't have the luxury to let things cool off for several days. Who starts writing projects that far ahead? But, if we're convinced that some cooling–off will improve the finished product, we can surely start on that writing project a little earlier so that we can at least break for lunch before beginning the editing and proofreading process.

Tips for the Task

One could never really get to the end of a list of editing tips, but here are some of the most common tips for the task.

1. *Give Sentences Strong Impact.* Choose nouns and verbs that create clear pictures in your readers' minds. Instead of saying, "The material had a peculiar odor," say that the material *stank,* or *smelled rotten* or *sweet* or *sour* or *rancid* or *pungent.* Better yet, be specific. Instead of using the vague term *material,* choose a term that's more descriptive like *fabric* or *wood* or, clearer still, *silk* or *oak* or even *green, slimy substance* if you don't know what the material is.

2. *Avoid Weak Beginnings.* The strongest words in a sentence are the nouns and verbs. Most sentences begin with nouns as subjects, followed by verbs and then direct objects.

> *Example:* John drives a fork lift.

Obviously, one wouldn't want every sentence written this way, or we'd soon be back to "Spot likes Puff." However, the construction does provide strong impact, and with the addition of modifiers to provide some variety, it is still the most forceful construction.

Weak sentence construction is often the result of a weak beginning. The most common offenders are the expressions "there is/are" and "here is/are." Such expressions lend no substance to the sentence and delay the message. But it's easy to get rid of them. Just cross them out and see what's left. Oftentimes it will be nothing, pointing up the emptiness of the sentence. For example, in the sentence, "There are many points to consider," we're not given any substantive information. In many cases the substantive information can be found in the next sentence.

Example: There are many points to consider. Two of these
are cost and training.

Combining the two sentences results in a much stronger sentence: "Many points need to be considered, including cost and training." Edit it and you have, "Cost and training need to be considered." Better yet, eliminate the passive construction and you have, "We need to consider cost and training."

At other times, cutting out the beginning "there is/are" costs you nothing and the remaining sentence will have stronger impact. For example, in the sentence "There are many people who disagree," if you take out *there are* and do a little editing, you end up with "Many people disagree."

3. *Avoid Redundancies.* Repetition, or redundancy, adds unnecessary padding to your writing as the following list exemplifies:

absolutely essential	gather together
actual experience	hard to the touch
adequate enough	join together
advance forward	mix together
advance warning	new beginning
aluminum metal	open up
assemble together	optimum best
bad problem	past history
basic fundamentals	period of time
basic necessities	personal opinion
big in size	plan ahead
bitter in taste	point in time
circle around	rectangular in shape
close off	red in color
close proximity	refer back
collaborate together	repeat again
collect together	serious danger
completely eliminate	short duration
connect together	single unit
consensus of opinion	sufficient enough
desirable benefits	surrounded on all sides
disappear from sight	true facts
end result	twenty in number
few in number	ultimate end
final conclusion	very unique
final outcome	visible to the eye

This list could run for pages, but these examples should make you aware of the problem so that you can check for redundancies in your writing.

4. *Avoid Wordiness.* One logical (or illogical) step beyond a redundant expression is an excessively wordy expression.

Wordy Expression	*Pared Version*
in this case	here
in most cases	usually
in all cases	always
in many cases	often
at the present time } at this point in time }	now
in order to	to
in reference to	about
in the very near future	soon
in the event of	if
involve the necessity of	require
in spite of the fact that	although
in the month of June	in June
in the neighborhood of	approximately, near
come to an end	end, finish
conduct an investigation of	investigate
previous to	before
it is necessary for us to	we must
of small length	short (or specify measurement)
afford an opportunity	allow, enable
as a general rule	generally
by means of	by
came into contact with	met
costs the sum of	costs
despite the fact that	though
during the time } during which time }	while
enclosed herewith } enclosed please find }	here is, we enclose
every effort will be made	we can, we will try
feel free to	please
first of all	first
for a total of } for the period of } for the reason that }	for
fully cognizant of	aware
give consideration to	consider
have need for	need
in addition to the above	also
in all probability	probably
in view of the fact that	because, since

make a decision	decide
of a confidential nature	confidential
of the opinion that	believe
on the occasion of	on, when
until such time as	until
with reference to	about

5. *Choose Simple Words.* Why choose simple words? You've been to school. Your readers have been to school. Why insult them with a simple vocabulary when you can impress them with big-time words, the ones you've been saving for just such an occasion? The answer to the question is a maxim: No one has ever been insulted by writing that is too clear. The K-I-S-S maxim (Keep It Simple, Stupid) applies here as well. If you want your writing to communicate and you don't want to make your readers work for the message, then you'll couch your message in the simplest vocabulary possible.

No need to impress. Remember that the need is to express. Save complexity and inflated language for those places where it can't be reduced. Below are examples of complex words paired with their simplified companions.

Complex	*Simple*
activate	start
advise	tell
aforementioned	previous
anticipate	expect
ascertain	learn, find out
assist	help
compensation	pay
disclose	show, tell
endeavor	try
execute	do
facilitate	ease, simplify
implement	begin, carry out
initial	first
initiate	begin
investigate	study
prior to	before
subsequent to	after
sufficient	enough
terminate	end
utilize	use
vehicle	car, train, bus, truck

Using vocabulary from the left-hand column, you might end up with something like this: Prior to our initiation of the aforementioned procedure and subsequent to our anticipated approval of the termination of development, we have executed our decision to implement the functions required.

Sounds like English, but what does it communicate?

6. *Choose the Concrete Word.* Words are symbols of things and ideas. Good writers choose words that create clear pictures in the reader's mind because the words symbolize specific things and ideas. The clearer the words, the clearer the picture. The less possibility there is for misinterpretation, the more likely it is that the reader will get the message the writer intended to send.

It's the vagueness of words like *facility* and *function* and *application* and *environment* that makes them poor choices. Such words don't create clear pictures in the reader's mind. If you read "The facility is impressive," what do you picture? A factory, a swimming pool, a cafeteria, a bathroom? Because they make the reader work needlessly to uncover the meaning, vague terms are time wasters and can cause the harried executive to fling your carefully documented report into the nearest wastebasket.

Sometimes a writer uses vague, abstract words like

capability
device
function
parameter
interface
analysis
area

in an effort to make the work behind the words seem more impressive. A real-world example points up the tendency: "Based on an extensive economic and functional capability survey of hotel sites around the country, the Marriott Hotel in New Orleans was again chosen for next year's meeting." Translated into plain English, the sentence probably should read, "After checking with several hotels to compare rates and meeting space, I again chose the Marriott Hotel in New Orleans for next year's meeting." The revised sentence sure takes some of the wind (and hot air) out of the writer's sails, but, after all, facts are facts.

7. *Avoid Dangling Modifiers.* A group of words that modifies a noun or verb needs to be placed as close to that noun or verb as possible or the modifier might end up dangling. In the sentence, "Walking home at dusk, the sunset was magnificent," can you tell who is walking home? The way the sentence is constructed here, the sunset appears to be doing the walking because "walking home at dusk" modifies the first noun it comes to: *sunset.* To correct the error, you need to insert the person who is walking. Two possible corrections are

> While I was walking home at dusk, the sunset was magnificent.

Walking home at dusk, I admired the magnificent sunset.

Maybe you didn't have a problem spotting that error. But look at this next sentence: "Hearing the news of the promotion, my spirits lifted." One's spirits cannot hear. The sentence needs a person doing the hearing. A correction might be

Hearing the news of the promotion lifted my spirits.

Another problem that leads to dangling modifiers is the pronoun *it,* as seen in the following example:

After observing the repair crews at work, it became obvious that they were reluctant to talk to us.

The introductory modifier—"after observing the repair crews at work"—is left dangling. The phrase appears to modify *it,* but *it* can't observe. Again, this problem can be corrected by putting a person in the sentence:

After observing the repair crews at work, we realized that they were reluctant to talk to us.

You'll want to be on the lookout for dangling modifiers like these. They're subtle but significant contributors to unclear communication.

8. *Avoid Noun Clusters.* Noun clusters are groups of nouns linked together as a unit. For example: joint fatigue life-enhancing fastener system development. The problem with such constructions is that some of the nouns in the cluster act as adjectives and some act as nouns, but the reader often cannot tell which are which. The more nouns there are, the worse the problem gets. The problem is compounded by the use of vague, abstract nouns such as the following examples taken from reports:

flexibility contingency option
management function hardware projection
state-of-the-art non-destructive evaluation and inspection procedures
construction, simplicity, economics, and availability standpoint
prime focus feed reflector configuration
tool surface preparation activity
mold release agent technology

In full sentences, as the following examples show, the situation does not improve.

> This person shall be responsible for the operation of the education programs support services.

> The Atlanta division has notified us of an unauthorized program to provide for a computerized action item system procedure.

> An artist's conception of a possible multi-directional tape laying machine configuration is presented.

In and of themselves, noun clusters are not total barriers to communication. When you stop to decipher them, you can usually figure out what you *think* the writer meant. But the problem is like measles. When one spot appears, lots more are sure to follow. If you're prone to the illness, take heart. It can be cured with lots of editing medicine.

9. *Avoid Prepositional Phrase Strings.* The prepositional phrase string is first cousin to the noun cluster. A prepositional phrase is a group of words beginning with a preposition (*of, by, to, for, on, under,* etc.) and ending with a noun or pronoun. The phrase acts as a modifier in a sentence. Prepositional phrases are helpful for clarifying "what kind," "how much," "when," and so forth. When strung together, however, they can confuse readers. Sometimes readers can't tell what modifies what. At other times, readers are forced to retain a lot of information contained in prepositional phrase strings until they can be attached to the word they modify. The following examples from business and industry demonstrate the condition:

> This center was the by-product of the analysis accomplished in the preparation of the requirements for the system review document reconfirmed in the other documents.

> Lack of positive action is evidenced by the absence of comments concerning action to be taken to prevent recurrence of cited deficiencies in responding to reports.

> Although the present program is of considerable assistance to us in ascertaining the overall effectiveness of our organization in the accomplishment of our mission, we hope that the enclosed comments will be helpful to you in further improving your understanding of the program.

When Abe Lincoln, the writer of the Gettysburg Address, wrote, "Of the people, by the people, for the people," he was using prepositional phrase strings, but the meaning was clear. In current usage, when writers use prepositional phrase strings, as shown above, they are stringing the reader along. The result is loss of clarity . . . not to mention conciseness.

10. *Use Parallel Structure.* Since much good technical and business writing contains lists, parallel structure is an important grammatical device to

master. (See p. 39.) Parallel structure comes into play when any words or groups of words appear in a series. In the sentence, "Discussion centered on the interpretation of specific provisions of the memo, the assignment of duties of the class coordinator, and to develop a plan of action," the third group of words is not structured in the same way as the first two. The lack of parallelism throws readers off and forces them to do what the writer should have done: change *to develop* into *development* to make the structure of the last phrase match the structure of the other phrases.

The items in the series could fall within a sentence as in the example above or they could be broken into a list as in the following example:

> Planning Control has the following responsibilities:
>
> [verb] 1. compiles all paperwork
>
> [verb] 2. distributes all paperwork to a central location
>
> [noun] 3. computerized planning system is used extensively
>
> [verb] 4. audits all schedules, tool orders, and reports
>
> [noun] 5. audits are done manually and by computer

In this example, the first, second, and fourth items in the list are parallel with each other and the third and fifth are parallel with each other, but the two groups are not consistently parallel throughout the list. To make the list parallel you must change them

> *computerized planning system* to
>   *uses computerized planning system extensively*
>
> and *audits are done manually* to
>   *audits manually and by computer.*

Of course, you could change all the items in the list to nouns, but that would require more work since only two of the five items are nouns.

Since lists are so common in technical and business writing, you'll need to check for parallelism when editing.

The Sargasso Sea—An Editing Challenge

The Sargasso Sea is a large, unusually calm part of the North Atlantic Ocean where seaweed and garbage collect. I end this chapter with a Sargasso Sea of unedited garbage, collected from technical and business reports. I have grouped them into general problem areas, but as you will see, many overlap, lending credence to the argument that where one type of writing problem surfaces, many others also collect.

FENCE STRADDLING

1. We cured the problem slightly.

2. These contractors will train a total of approximately three hundred students.

WORD CHOICE (INFLATED LANGUAGE)

3. The agency will plan and oversee the execution.

4. Since implementation of the revised policies and procedures is being effected prior to the republication of this document, an early reply would be appreciated.

5. Remove all exogenous materials not indigenous to the tool.

6. Minimize error conditions associated with incompatible load situations.

7. Illumination is required to be extinguished on these premises on the termination of daily activities.

8. Management participation should aid in the training process because it also hastens acceptance because subsequent efforts will be predicated upon sustenance of their own prognostications rather than anticipation of failure.

9. This system, in effect, represents computer power at its best to generate radical productivity improvements through the application of a global view to optimize individual manufacturing components.

10. To maximize performance from this molding technique, characterization of the coefficients of thermal expansion of the constituent material, i.e., the rubber mandrels and the mold cage, must be controlled to insure dimensional stability with respect to engineering tolerances.

FAILURE TO PASS THE "SO-WHAT?" TEST

11. Section ten provides an outlet for identifying requirements of a temporary nature on those that have a limited effectivity. No such requirements were identified, and so section ten has been eliminated.

12. The technical training facility established on site by the Chicago office is utilized in teaching technical training courses.

13. Note: Inherent in the term "reusable" is some period of longevity, as the bag can be used over and over again.

14. The plant makes no primary structures. Primarily it makes secondary structures.

15. The team is drafting a two-year study and will not submit a final report until completed.

CLEAN-UP

Possible revisions to the sentences above are as follows:

1. We cured the problem.

2. These contractors will train approximately 300 students.

3. The agency will execute the plan.
   (or)
   The agency will carry out the plan.

4. Because the revised procedure will be used before the document is republished, please reply soon.

5. Remove all foreign matter from the tool.
   (or)
   Clean the tool.

6. Reduce errors resulting from incompatible loads.
   (or)
   Don't overload the system.

7. Turn the lights off when you leave work.

8. Management participation in training will hasten worker acceptance.
   (or)
   For best results, managers ought to be involved in training.

9. This computer-based system improves productivity because it takes a broad view of manufacturing to increase each component's effectiveness.
   (or)
   This computer system quickly supplies the best data for the most efficient manufacturing operation.

10. For the most efficient molding technique, expansion of the rubber mandrels and the mold cage must be controlled to ensure that the

part is made within engineering tolerances.
(or)
To best use this technique, the coefficients of thermal expansion of the rubber mandrel and the mold cage must be controlled within engineering tolerances.

11. Section ten, designed to identify temporary requirements, was eliminated when none were found.

12. Technical training courses are offered at the Chicago office.

13. [This sentence is unnecessary since the meaning *reusable* is obvious.]

14. The plant makes mainly secondary structures.

15. Upon completion of a two-year study, the team will submit its final report.
(or)
The final report will be completed at the end of a two-year study.

More of these choice tidbits follow in the next chapter.

7
Cases
In Point

"**D**o as I say, not as I do," seems an apt adage if one looks at much of the business and technical writing produced today. The object of this chapter is to demonstrate, by example, just how widespread poor communication is in business and industry. I hope that by entertaining you with others' mistakes I'll also inspire you to rise above the commonplace to the higher reaches of true communication.

Perpetuators of the problem of poor communication come from all walks of life. To classify the examples in this chapter, I have divided them into four categories:

1. Those who should know better—people who write for a living

2. Those who do know better—people who have written spoofs of bad writing to point up the problem

3. Those who know not what they do—people who make blunders without realizing it

4. Those who cannot be saved—people whose writing is so undecipherable that one wouldn't even know where to begin to fix it

Those Who Should Know Better

This group comprises professional technical writers, experts publishing in business and technical writing journals, professors of technical and business communication, and the like. One would think that these people, surely, could show us the way to clear writing by example. But, alas, such is not always the case. A few examples should prove the point.

In a journal on business communication, one writer states, "I cannot emphasize strongly enough the need to credential yourself heavily as an expert in your chosen field. Credentialing is not simply a matter of degrees." Then what *is* it, you might ask, since the author has coined a new word, *credentialing,* and has converted the noun *credential* to a verb.

A promotional brochure for a leadership seminar describes ways to help participants with "conflictual problems." Again, we are introduced to a new word—*conflictual*—of uncertain meaning.

A technical writer describes the success of the team concept in his organization as follows: "The effectiveness of this integrated, functional and projectized organizational approach has been proven to be very successful." Not only does he coin the word *projectized,* but he adds it to *integrated, functional,* and *organizational*—all to modify *approach.* Put 'em together and what have you got? To top it off, the writer adds a passive construction—*has been proven to be.* The sentence could, and should, read something like this: "The team approach works" (or, better yet) "Teams work."

In a seminar offered by The Council Of Communications Societies(!), the program describes the following activity scheduled for 10:20–10:45 A.M.—"Coffee and Conversion." Conversion to what, I wonder.

And finally we have the following excerpt from the Profile Disk Drive manual for Apple computers:

> *Your Profile Drive is packed in a cardboard shipping carton. After you open the carton, remove the top layer of thick, foam material and you will find a small cardboard box lying on top of the drive. The box contains this manual.*[9]

Those Who Do Know Better

This group contains spoofs of bad technical writing style. If the writer knows enough to make fun of the bad technical writing that abounds, he or she can probably rise above it.

The first comes from the military establishment.

> *A COLONEL ISSUED THE FOLLOWING DIRECTIVE TO HIS EXECUTIVE OFFICER:* "Tomorrow evening at approximately

2000 hours, Halley's Comet will be visible in this area, an event which occurs only once every 75 years. Have the men fall out in the Battalion Area in fatigues, and I will explain this rare phenomenon to them. In case of rain, we will not be able to see anything, so assemble the men in the theater, and I will show them films on it."

EXECUTIVE OFFICER TO COMPANY COMMANDER: "By order of the Colonel, tomorrow at 2000 hours, Halley's Comet will appear above the Battalion Area. If it rains, fall the men out in fatigues. Then march to the theater where the rare phenomenon will take place, something which occurs only once every 75 years."

COMPANY COMMANDER TO LIEUTENANT: "By order of the Colonel in fatigues, at 2000 hours tomorrow evening, the phenomenal Halley's Comet will appear in the theater. In case of rain in the Battalion Area, the Colonel will give another order, something which occurs only once every 75 years."

LIEUTENANT TO SERGEANT: "Tomorrow at 2000 hours, the Colonel will appear in the theater with Halley's Comet, something which happens every 75 years. If it rains, the Colonel will order the Comet into the Battalion Area."

SERGEANT TO SQUAD: "When it rains tomorrow at 2000 hours, the phenomenal 75-year-old General Halley, accompanied by the Colonel, will drive his Comet through the Battalion Area theater in fatigues."

The next group includes famous quotations from history translated into gobbledygook. Try your hand at translating them back into the original.

FAMOUS QUOTATIONS FROM HISTORY — TRANSLATED INTO GOBBLEDYGOOK

1. It is regretted that this speaker lacks a multiplicity of lives which, under prevailing circumstances, might be offered on behalf of his nation.

   TRANSLATION:

   _____

   _____

2. An ultimate end to corporeal existence is preferred to continued viability without the attendant liberties generally associated with the rights and privileges of a free people.

    TRANSLATION:

    _____

    _____

3. After assuring yourself that all pertinent procedures and preparations have been accomplished, permission is granted to initiate the overall implementation of combat operations.

    TRANSLATION:

    _____

    _____

4. Hear Ye! This is classified information on a need-to-know basis. British, redcoated, armed, are proceeding in this direction.

    TRANSLATION:

    _____

    _____

5. The full combat potentials available to me have not been effectuated at this point in time.

    TRANSLATION:

    _____

    _____

6. Defensive fire operations will commence only after it is possible to discern the distal corneas, surrounding the pupils, of the advancing enemy.

    TRANSLATION:

    _____

    _____

7. At some unspecified point in time, this speaker assures his certain reversion to this place.

    TRANSLATION:

    _____

    _____

The translations are as follows:

1. I regret that I have but one life to give to my country.
   (Nathan Hale)

2. Give me liberty or give me death.
   (Patrick Henry)

3. Fire when ready.
   (Admiral Dewey)

4. The British are coming!
   (Paul Revere)

5. I have not yet begun to fight.
   (John Paul Jones)

6. Don't fire until you see the whites of their eyes.
   (William Prescott, Commander at Bunker Hill; also attributed to Israel Putnam)

7. I shall return.
   (Douglas MacArthur)

In a similar vein, an instructor asked her students to take good, clean prose and translate it into jargon; then, to do the opposite to a piece of jargon. A few examples follow.

The Original:
> *Call me Ishmael. Some years ago—never mind how long precisely—having little or no money in my purse, and nothing particular to interest me on shore, I thought I would sail about a little and see the watery part of the world.*
> *—Herman Melville,* Moby Dick[10]

In Jargon:
> *You may identify me by the nomenclature of Ishmael. At a point in time several years previous to the current temporal zone—the precise number of which is extraneous information—devoid of sufficient monetary resources and lacking physical and/or psychical stimuli within the confines of my sphere of activity on land, I initiated several thought processes and concluded that I would commandeer a vessel of navigation with which to explore the aquatic component of this planet.*
> *—Vicki Hunter '81*

---

[10] Debra Shore, "Identify Me by the Nomenclature of Ishmael: Undergraduate Writers Interface with Jargon," *The Brown Alumni Monthly,* 81, No. 5 (Feb. 1981) 20–23.

In Jargon:

> *At the inception of the primary fabrication time-phase, when the penultimate intelligence unit synthesized the geophysical locus and its concomitant gaseous hyperterran coordinates, said mineral consolidation region failed to possess proper substance and volume reference points and displayed a lack of wave frequency vibrations in the specific imposition registers over the anterior surfaces of the geotropic fault formation and with the aqueous gas-to-firmament interface destabilized by a major kineticizing meteorological manifestation. The previously discussed entitical sentence unit expressed a desire for increased wave frequency modulations and with the immediately subsequent amplification adjustment, registered his positive reactions.*
>
> *—Mack Reed '81*

The Original:

> *In the beginning, God created heaven and earth.*
> *The earth was without form, and void; and darkness was upon the face of the deep. And the Spirit of God moved upon the face of the waters.*
> *God said, Let there be light; and there was light.*
> *And God saw the light that it was good; and God divided the light from the darkness.*
>
> *—Genesis 1:1-4*

Those Who Know Not What They Do

This group consists of those who do not say what they mean or are misunderstood by their readers because they think they say one thing and their readers think they say another.

Sometimes a lack of clarity is rewarded, as an article called "Creative Obfuscation"[11] points out. The author reports that gobbledygook is rewarded when readers or listeners think it's coming from experts. He cites several examples:

---

[11] O. J. Scott Armstrong, "Creative Obfuscation," *Chemtech,* May 1981, pp. 262–264.

- An actor was hired to give a speech on a topic he knew nothing about to a group of well-educated professionals. His phony biography was impressive and he looked the part, but his talk was sheer gobbledygook, with just enough of the right jargon thrown in to sound convincing. The audience thought he gave a clear, stimulating lecture.

- Readers of management journals were surveyed to learn their assessment of articles appearing in the journals. Those with the most difficult readability were given the highest competency rating.

- A trade journal that had rejected a clearly written paper later accepted the same paper after the author rewrote it to the point of incomprehension.

While the point of the article may be sad but true, we should nonetheless strive for clear communication as the most likely to succeed most of the time. As the following examples from accident reports indicate, problems can arise if we fail to make our point perfectly clear. In each case, the writer says something other than what was intended.

> *I pulled away from the side of the road, glanced at my mother-in-law and headed over the embankment.*
>
> *I had been driving for 40 years when I fell asleep at the wheel and had an accident.*
>
> *To avoid hitting the bumper of the car in front, I struck the pedestrian.*
>
> *The pedestrian had no idea which direction to run, so I ran over him.*

Another example comes from a description of office help available through an employment service:

> *Harriet—versatile and efficient. Uses all keypunch equipment, any CRT with numbers at the top. Her dependability and expertise usually turn one–to–two–week assignments into one–to–three month situations.*

An announcement distributed to tenants of an apartment complex describes precautions to be taken during cold weather. It also expresses the following concern:

> *PLEASE NOTE: We would like to express our concern for you animals.*

Obviously, they mean *your* animals, but that's not what they wrote! Examples from newspapers point up similar bloopers:[12]

- From the Dallas *News:* *During the gala, the Bushes shared the spotlight with such Texas luminaries as Lynda Johnson Robb and former Governor John Connally and his wife, Nellie. The Connallys, in fact, generated the second-biggest ovation—behind the Bushes.*

- From the help-wanted section of the Bakersfield *Californian:* *Security officer with shoplifting experience . . .*

- From an ad in the Hurst, Texas, *Mid-Cities Daily News:* *Lost: small apricot poodle, reward, neutered like one of the family.*

Translations from a foreign language into English also pose problems, as the following examples show. The first is an advertisement in the English-language newspaper *Arab Gazette,* published in Saudi Arabia. As you can see, the poor translation garbles the message almost beyond comprehension in some places.

FOR HARD AND TROUBLESOME WORK
SOLID..FORCE..HIGH ABILIT Y IN THE PRODUCTION

- Capacity From:0,8 M^3 And Biger It Supplied With Arm For Dig a Cables And Water Pipes Excaration.. To Depth Mor Than 4M.
- Our Rates Is Rivalry The Spare Parts Abundant Conservation Is In a Work Places Befor And After Conservation Period.
- Guaranty For One Year.

COGEMA

COGEMA CIMAS

I.MANCO.Lo.l

RIYADH: SHOW ROOM: DHABAB STREET: TEL: 4910703
JEDDAH: ALAMOUDI BILDING . KILO 14 . MADINAH STREET
O.B.B.CO ALBAKRY TEL: 6830931

---

[12] "How's That Again?" *Reader's Digest,* December 1982, p. 138.

Another is from a help-wanted advertisement which appeared in an English-language newspaper in India. The advertisement read as follows:

> **We need a**
> —Scribbler
> —Who can ring bells
> —Temporary
> —Ladies will be preferred
> —Bring your certificates

Translation:

Wanted—Someone who takes shorthand and can type, for a temporary position. Women preferred. Bring letters of recommendation.

Since so much technical writing now involves translation into English, the translator has an even greater responsibility to express the message clearly and accurately.

Those Who Cannot Be Saved

Those in this group seem beyond help. The only hope for people in this category is to memorize this book and take every word to heart. If their numbers weren't so vast, one could ignore them. But, alas, that isn't possible since their misguided efforts at communication come from government, business, and industry—virtually all walks of life.

Take this first typical example from a government contract agreement:

> *It is agreed that, in the event the Government elects not to continue funding of an individual technology whenever the Estimate-To-Complete (ETC) of the technology exceeds the initial estimated Total Item Amount (All Phases) as set forth on the AFSC Form 705s (705s) or, in the event the technology is modified pursuant to the "Changes" clause of the General Provisions hereof, the modified Total Item Amount, by twenty-five percent (25%) or fifty-thousand dollars ($50,000.00), whichever is greater, and the Government, therefore, discontinues said technology pursuant to the "Changes" or "Termination" clauses of the General Provisions hereof, then the Contractor shall not be entitled to incentive payment.*

Not bad . . . only 108 words in the sentence!

A second example comes from *The Register—A Newsletter of the Appalachian Trail:*

> In a controversial recent decision, Forest Supervisor George Olson provided for a pre-relocation baseline information study conducted by an unbiased third party to be followed by three years of carefully-monitored hog relocation, on a put-and-take basis, in two new areas of the Nantahala N.F. [National Forest]. Should environmental impacts prove unacceptable, hog relocation would cease and efforts would be made to eliminate the hog population.[13]

Wading through all this hogwash, we come up with the following simpler translation:

> Forest Supervisor George Olson authorized an outside group to study the impact of releasing hogs into new areas of the Nantahala National Forest. After the study, hogs will be relocated for a trial period. If, after three years, the hogs have damaged the forest, they'll be removed [killed?].

The next three contributions come from the world of business. The first is a letter I received from the engraving company that makes my business cards. It doesn't instill me with confidence in the company's ability to get my cards right!

---

[13] *The Register—A Newsletter of the Appalachian Trail,* May 1985, p. 5.

wrong initial   *should be lower case*

October 1, 19__

Carol Ⓦ Barnum, Ⓗ D.   *wrong punctuation mark*
Southern Technical Institute
Marietta, Georgia 30060

Dear Ms. Barnum⃝   *fragment*

    We received your letter concerning the proof for your
business cards. Concerning the letters STI not lining up under
member. We have measured the lettering on this side and the
letters do line up under member. There is an optical illusion
but if you measure from the longest points the letters are in
line.

fragment   Also concerning the position on the cards. Your name and the
address will be in the same position on the cards as the first
order you received. If this is not satisfactory we will need to
make a new engraving for your cards and since this was not
mentioned in your letter when we made over the original order you
will be charged for another engraving.

    I have enclosed a copy of the first card we did for you with
the wrong spacing for you to see the position of your name and
the address. Please let me know if this is satisfactory.

                Yours Ⓣruly,

                Engraving Company

should be lower case

The following example comes from an employee who wanted to be considered for promotion. The employee was told to write a brief history of his employment with the company and the reasons why he should be promoted.

Mr. _____ join PYC Company in March of 1977 as a plant designer. After complating a very large belt conveyor contract was promoted to senior plant designer. While Mr. _____ has been with us has proform his job in a professional and satisfactory manner. He assume responsibility will and work well with know or very little supervision. He takes controll of problems willingly and bring them to a fast conclusion.

By the way, he was *not* granted the promotion. Obviously, technical competence was not the only criterion for promotion.

This next example was written by a person working in the clinical lab of a hospital. The letter is in response to a telephone inquiry. See if you can decipher its message . . . and remember, it went out on letterhead stationery!

Denise _____ , Customer Service:

In reference to the problem that we discusses early over the phone, I'm sending you copies of the invocies inwhich I'm having problems matching invoices with the packing list. These items are from _____ , which I'm informed that _____ invoices you and in returned you invoice us.

Attached is one invoice and packing list in which I can reference a _____ number which corresponds with the invoice. If you can invoice us in this manner, the problem would be solved. Any suggestions or comments would be appreciated in helping me solved this problem.

Yours truly,

Clinical Lab

The point of the examples in this chapter is that bad grammar, bad form, lack of clarity, improper or inappropriate word choice, and countless other errors exemplified in these cases have no place in technical or business writing. They damage the writer's image, as well as the company's, and impede— or even prevent—communication.

8
The Proof of the Writing is in the Reading

Y ou've written your report, let it cool off, and then edited it. Great. But you're not done yet. In fact, what's left—proofreading—could be the most important phase.

If your report is brilliant but full of spelling errors, if it sidesteps all the editing pitfalls but has typographical errors, if it represents hours of labor but goes to your readers with a line or even an occasional word or two left out, your report is incomplete, or worse, wrong. At the very least, readers who catch the errors will think less of you for sloppy work—but worse than that, they'll begin to doubt your credibility altogether if they see obvious errors you should have seen but didn't. Your report is largely judged by its appearance. If it has been scrutinized before making its debut, your readers, on seeing it, will assume that what they *can't* see—all the work that led to the writing of the report—was as carefully handled as what they *can* see.

Sometimes, you don't make your own best proofreader. It may be because you're too close to the report or too weary of it. In either case, it's best to get someone else to do the proofreading for you. Frequently, a good secretary can do it. A colleague familiar with your work can also be a good proofreader. You should still do the final proofreading yourself since you can never be sure that another person will catch everything; and after all, it's your name on the report, not someone else's. Moreover, two proofreaders provide insurance. One will catch some things, while the other will find different errors.

If a page has to be retyped, then you must proofread the whole page again unless it comes from a memory typewriter or word processor. On copy produced by these, you need only proofread the corrections. As word processors become even more common, the job of proofreading will become easier. Without the aid of the word processor, however, the byword of proofreading is to read and re-read.

Proofreading Pointers

The following pointers will make your proofreading efforts more productive.

1. *Slow Down.* If you let your eyes skim over the page, you'll skip over a lot and consequently miss a lot. At high speed, everything will look fine. At slower speeds, you're more likely to see the misspelled word, the missing letter, the same word at the bottom of one page and the top of the next.

How do you slow down? By reading out loud. Silly as it sounds, it works. If you have to say the words, they have to be there. If it's not possible to read out loud because it would disturb your co-workers, then mouth the words as if you were actually saying them with the volume turned down. In that way, you'll still maintain the reading speed that lets you see what's actually on the page. Better yet, find a room to proofread aloud with another person reading along. The two of you will be listening and reading together, providing greater assurance of catching mistakes.

2. *Read Backwards.* Not everyone likes this technique, but those who do swear by it. When you read backwards, you see each word in isolation because you're not reading in context. Obviously, with this method you're reading for spelling, not meaning. Since you can't slide over them, misspelled words will jump out at you.

3. *Check Hyphenation.* The dictionary breaks words into syllables. Hyphenation looks like this: *hy · phen · a · tion.* The dots indicate the "legal" breaks in the word: those places where you can hyphenate if you need to at the end of a line. The person typing your report will have to look up words that need to be divided at the end of the line in order to locate the appropriate breaks. If the typist doesn't do it, you will have to. It wouldn't hurt to check a couple of hyphenated words to make sure that they're being correctly divided.

Also, make sure that when you look up a word in the dictionary, you locate the exact word you want. For example, *phi • los • o • phy* is not broken the same as *phil • o • soph • i • cal* even though they both have the same root. Close is usually not good enough.

Another thing—don't leave one or two letters by themselves at the end or beginning of a line. Since it already takes one space for the hyphen, it's better to complete the word at the end of a line than to carry two letters over to the next line. Likewise, it's better to start a word on a new line rather than divide it after only one or two letters.

Thankfully, these typing rules soon may no longer be necessary since many word processors can make all the decisions about hyphenation—either through a data dictionary that indicates how to hyphenate or by "wrapping" a word that won't fit on a line down to the next line, avoiding the problem of hyphenation altogether. But until then . . . .

Some other rules of hyphenation are also useful to know.

- Hyphenate two words used as a single modifier.

    Examples: single-digit number

    well-documented article

    half-completed project

    Exception: Words joined by an adverb ending in -*ly* are not hyphenated.

    Examples: Frequently cited case

    normally hot day

Note: Check the dictionary for words that are normally hyphenated. The dictionary will show the hyphen.

- Hyphenate numbers in combination from twenty-one to ninety-nine.

Note: Do not, however, hyphenate numbers like six hundred sixty or two thousand twenty.
Do hyphenate numbers like two thousand twenty-two.

- Hyphenate words with prefixes like *ex-, anti-, pro-, self-,* or *non-.*

  Examples: anti-establishment

  ex-husband

  pro-military

  self-taught

Note: This rule may change in the near future as the trend is to join many of these constructions without a hyphen.

- Hyphenate words for clarity.

  Examples: re-cover the sofa (versus recover from an injury)

  a co-op student (versus a chicken coop)

4. *Check the Form of Numbers.* Have you written out the required numbers and used the proper Arabic numerals or fractions where required? The rules about numbers vary from source to source and company to company, so once again, the main rule is to be consistent. The following general guidelines are practiced.

- Spell out whole numbers to ten. Use Arabic numbers (16, 25, 37) after number ten. However, for a series of numbers—some under ten and some over ten—use Arabic numbers for all in the series.

  Examples: The meeting room seats nine people.

  The largest conference room seats 24 people.

  There will 6 people at the first meeting and 17 at the second.

- Spell out any number that begins a sentence.

  Example: Seventeen people will be at the meeting.

- When two numbers are used together, one to indicate quantity, the other to indicate type, spell out the first number or the shorter number.

  Examples: sixteen 4-inch bolts

  100 ten-inch bolts

- Spell out numbers that are rounded off.

  Examples: about five hundred people

  approximately forty thousand pieces

- Use Arabic numbers for exact measurement and with abbreviations and symbols.

  Examples: 5 mm

  16 bbl

  3%

  6 1/2

  6.25

  9′–12′

  20°–30°F

  $70–$90

5. *Check for Proper Capitalization.* In the old days, people used to capitalize all important words for emphasis.

  Example: This is Very Important!

Then the practice died out. Now, unfortunately and unofficially, the practice is creeping back into a lot of technical and business writing. The best practice is to capitalize as little as possible. When in doubt, leave the capital letter out. The general rules for appropriate capitalization follow.

- Capitalize all proper names of people, places, and things.

  Examples: Washington Monument

  George Washington

  Constitution Avenue

  Potomac River

  Sahara Desert

- Capitalize honorary and official titles.

  Examples: Chairman Johnson

  President Lincoln

- Capitalize the days of the week, months, holidays, holy days.

  Examples: Sunday

  March

  Thanksgiving

  Palm Sunday

- Capitalize acronyms and abbreviations of proper names.

  USA (United States of America)

  STC (Society for Technical Communication)

  BASIC (Beginner's All-Purpose Symbolic Instruction Code)

- Do *not* capitalize the following:

  Directions—southern, northwestern (*do* capitalize regions like the South)

  Seasons—spring, summer, fall, winter

  General titles—senator, president, department head

6. *Check Headings.* When you proofread, you will probably have a tendency to skip over the headings and head for the paragraphs. Don't. The headings must be checked for spelling and consistency. Are all the words there? Are they properly spelled? Are all headings of the same level presented in the same format? Are all the appropriate headings underlined? Do the third-order headings have the appropriate end punctuation?

7. *Check Tables and Figures.* Are they all there? Are they where they belong in relation to the text? Are they positioned correctly? Are they listed correctly on the illustrations page with the correct page numbers?

8. *Check References.* Have you quoted accurately? Have you footnoted correctly and consistently? Are the page references accurate? Is the list of sources accurate and complete?

9. *Check Page Numbers.* Are the pages in the right order? Are the numbers located in the same place on each page? Are the pages with illustrations numbered correctly? Are all the pages right side up? Are all the pages there?

10. *Check Names.* The spelling of proper nouns is terribly important, especially the names of individuals and companies. Nobody likes to see his or her name or the name of the organization misspelled. A misspelled name is a sure source of embarrassment to you. Since you won't find most proper nouns in the dictionary, you'll have to check other sources to make sure the names are correctly spelled. Remember, if a person sees his or her name misspelled, that person is likely to think you've done sloppy work throughout your report.

Quotable Misquotes

Some of my favorite proofreading blunders (garnered from my students' papers, which were all supposedly proofread) include the following:

- personnel manger [another variant of *personal manager*]

- crudentials

- An engineer has to have a large vocabulary because he is constantly encountering very important and suffocated people.

- The fast forward and rewind positions will move the tape forward or backward, respectfully.

- Lowering the thermostat at night can be a chilling experience in the mourning.

- He has to do it wether he likes it or not. [When I questioned the student about the spelling of the word *wether,* he said that he had looked it up in the dictionary. So I did, too. Sure enough, it's there. Only it means "castrated male sheep"!]

9
Diction's Dirty Dozen

What *is* diction? Word choice. The correct word for the situation. The difference between incorrect and correct usage. Do you want to *lie* down or *lay* down? Do you want to see the *principal* or the *principle*? Did you order *stationary* or *stationery*? That's diction, and the following "dirty dozen" are the worst offenders.

1. Affect-Effect

Affect is a verb. It means *to influence.*

> *Example:* The 55 mile-per-hour speed limit *affects* everyone.

Effect is a noun. It means *result.*

> *Example:* The *effects* of the 55 mile-per-hour speed limit are fewer highway deaths and reduced fuel consumption.

Affect and *effect* sound pretty much the same and, as you can see from the examples, can be used in the same discussion. But they're altogether different. One is a verb; the other a noun.

Now, you could stop there, and you'd probably be in good shape. However, for the adventuresome, there's one more usage for *effect,* which is not as well known as the noun usage. *Effect* can also be a verb. It means *to bring about* or *to result in.*

Example: Please *effect* his transfer to another department.

If the verb usage for *effect* causes you problems, you can easily live the rest of your life without it. The most common misuse involves *affect,* the verb, and *effect,* the noun.

2. Among-Between

Between refers to two things or people.

Example: He sat *between* John and Martha.

Among refers to three or more.

Examples: She sat *among* the thirty students.

He can distinguish *among* several types of machines.

3. Amount-Number

Amount is used for singular units: *milk, money, work.*

Examples: The *amount* of money is large.

The *amount* of milk is small.

The *amount* of work is staggering.

Number is used for quantities expressed in units indicating the plural: *drops, coins, hours.*

Examples: The *number* of drops is small.

The *number* of coins is large.

The *number* of hours is reasonable.

Note: The same situation applies with *less* and *few* (or *fewer*). Use *less* with a singular unit and *few* with plural units.

> *Example:* I have *less* money and *fewer* dollars since I quit my job.

4. Bad-Badly

Bad is an adjective. It modifies a noun.

> *Example:* He was a *bad* boy.

Badly is an adverb. It modifies a verb, an adjective, or another adverb.

> *Example:* I did *badly* on the exam.

Simple enough. But what sorts of modifiers do you use with verbs like *seem, appear, feel, taste, smell, become, sound*? These are called linking verbs because they link the subject with the object. In the sentence "I feel (*bad* or *badly*) about the accident," the correct response is *bad* because the verb (*feel*) links the subject (*I*) with the adjective (*bad*).

Many people want to say, "I feel *badly* about the accident," thinking they need an adverb to modify the verb. But, if you think about what linking verbs do and what the sentence really means, if you use *badly* to modify *feel,* you'll see why it doesn't work. What you may mean when you write "I feel *badly*" is that you have poor tactile sensation in your fingertips! You can feel *strongly* about something, or you can feel *strong* because you ate your spinach. The choice of an adverb or adjective can change the meaning of the sentence.

5. Comprise-Consist

The rule is easy, but it's hardly ever applied correctly. *Comprise* means *to consist of, to contain,* or *to include,* but it cannot always be used in place of these expressions. It can only be used in expressions that mean a division of the whole such as "the whole *comprises* the parts," to mean the whole "contains" or "consists of" the parts. It cannot be used to mean the reverse, as in "the parts *comprise* the whole."

You can say, "The subject *comprises* many topics," but you cannot say, "Many topics *comprise* the subject" or "The subject *is comprised* of many topics." In these cases you must choose another approach like "The subject *consists* of many topics."

Note: Like many rules, this one is changing after suffering repeated abuse from sources who should know better. So, if the "official" use still gives you trouble, steer clear of *comprise*; and the rule will probably change in the not-too-distant future.

6. Due To-Because Of

Many people say *due to* when they mean *because of.* The problem is *due to* a lack of familiarity with the distinction between the two expressions. (*Due to* is used correctly in the preceding sentence.) *Due to* is used *incorrectly* in the following sentence: "The problem results *due to* a lack of awareness . . ."

If you see the distinction in the use of the expression in the two examples, move on to #7. If you don't, the following explanation should clear things up. *Due to* can only follow a linking verb (either the *to be* verb or a verb like *seem, appear, become, smell,* etc. See #4 for further explanation of linking verbs). You can say, "My late arrival was *due to* the traffic jam." But you can't say, "I arrived late *due to* the traffic jam." In the second sentence, you have to use *because of* rather than *due to.* Likewise, you can't say, "*Due to* the traffic, I was late." *Due to* can only *follow* a linking verb.

7. Farther-Further

Farther is for geographic distance.

Example: I won't go a step *farther.*

Further denotes additional quantity, extent, or degree. (Think of it in reference to ideas.)

Example: He has one *further* point to make.

P.S. This is another rule in the process of transition, but until *further* notice, you are advised to keep the meanings separate and distinct.

8. Kind-Kinds

Kind and *kinds* have distinct uses based on number.

Kind is singular.

Example: That's my *kind* of play.

Kinds is plural.

Example: Those are my *kinds* of plays.

Their favorite companions (also incorrectly matched up much of the time) are *this* or *that* and *these* or *those*.

This and *that* are singular (*this* for objects near at hand; *that* for objects farther away). *This* and *that* go with *kind*. *These* and *those* are the corresponding plurals. They go with *kinds*.

Note: First cousin to *kind* is *type* and its plural, *types*. The rules for these words are the same as those for *kind* and *kinds*. *Type*, however, has been serving time with other words, generally resulting in a bad construction, usually hyphenated.

Examples: college-type atmosphere

bologna slicer-type device

People must think the word *type* lends importance to a common construction like *college atmosphere*. It doesn't. It just adds bulk.

One last caution about the use of *type*. It is incorrect to write, "He is a technical type person." It is correct to write, "He is a technical type of person." *Type* (and *kind*) must be followed by the preposition *of*.

9. Lead-Led

These look like a pair, but they're actually three words. That's what causes much of the confusion, since two of the three sound the same but are spelled differently, and the third is spelled like one of the others but is pronounced differently.

The three are as follows:

lead (pronounced with a long *e* as in *feed*).

> *Example:* You can *lead* a horse to water but cannot make it drink.

led (pronounced with a short *e* as in *fed*). It is the past tense of the verb *to lead.*

> *Example:* Yesterday, you *led* a horse to water but could not make it drink.

lead (pronounced with a short *e* as in *led*). Now we're back to the same spelling as the first word, but this time it's with a different pronunciation — thus, the confusion. This one is a noun, meaning the metal or the element.

> *Example:* This paint contains no *lead.*

10. Lie-Lay

This pair and its relatives — *sit-set, rise-raise* — cause many problems. But, if you learn the rule for one, you'll know it for all three.

lie/lay   sit/set   rise/raise

The word on the left of the slash is something that *you* do. The word on the right of the slash is something that you do *to* something. It requires a direct object, the *thing* you're doing something to.

For instance: But:

You *lie* down   You *lay* your body down.

You *sit* down.   You *set* yourself down.

You *rise* at 7 A.M.   You *raise* the window.

(And, by the way, you don't *raise* children; you *rear* them. But that's another subject altogether!)

11. Lose-Loose

This is mostly a problem of pronunciation and spelling.

Lose is *not* pronounced like its look-alikes *close* or *dose* (which are pronounced differently from each other!) It is pronounced like *snooze*.

Loose is pronounced like *goose.*

> *Examples:* You *lose* money.
>
> It is *loose* in your pocket.

The noun that corresponds with the verb *lose* is, of course, *loss.*

12. Principle-Principal

Again we have two words that sound the same but have different meanings and pronunciations.

Principle means a *rule* or *truth.* It is always a noun.

> *Examples:* I stand on my *principles.*
>
> It is the *principle* of the matter.

Principal can have several meanings.

It can be an adjective meaning *main* or *chief.*

> *Example:* His *principal* source of support is from stocks.

It can be a noun meaning *head of an organization.*

Example: The *principal* addressed the student body.

(Do you remember the ditty from school that went, "The *principal* is my *pal*"? It helps you remember the spelling.)

It can be a noun meaning *a sum of money.*

Example: The *principal* plus interest is due in six months.

To simplify matters, remember that *principle* means rule or truth and *principal* means everything else.

Want to make it a baker's dozen? If so, here's one more.

13. Stationary-Stationery

This one also has a grade school ditty that goes with it. The *–er* in pap*er* goes with the *—er* in station*ery*. The "other" means motionless, standing still.

Example: The *stationery* is *stationary* since there is no wind.

P.S. Course and *coarse* cause similar problems. But I've used up my baker's dozen.

10
Punctuation
Pointers

My purpose in this chapter is not to take you through the whole punctuation routine one more time. Rather, I will point out the basic rules for the most commonly used punctuation marks in a greatly reduced, easy-to-use form. You'll need to know a few terms, which I'll give you along the way. While I may leave certain points unaddressed, the following presents what I consider the pressure points of punctuation—based on the errors I've seen time and again in business, industry, and the college classroom.

A Few Ground Rules

We need to begin with a few ground rules about the sentence to know how to punctuate one. Choices about commas, semicolons, and colons are largely dependent on the kind of sentence you write.

There are four types of sentences:

1. *The simple sentence* contains a subject and verb that can stand alone (that is, it is a complete thought).

   Example: John loves grammar.

2. *The compound sentence* contains a subject and verb that can stand alone followed by a "connector" and then another subject and verb that can stand alone. The connectors are the following: *and, or, but, nor, for* (and sometimes *so* and *yet*). Think of it as two simple sentences joined by a connector.

   Example: John loves grammar, but he bores his friends with
   his incessant talk about it.

3. *The complex sentence* contains a subject and verb that can stand alone and a subject and verb that cannot.

> *Example:* Since he was late again, he was fired.

4. *The compound-complex sentence* is a compound sentence with at least one subject and verb that cannot stand alone. Think of it as a combination of the compound and complex sentence, as the name implies.

> *Example:* Since he was late again, he was fired, and he had
> to leave without saying goodbye.

TIME OUT FOR DEFINITIONS

A group of words containing a subject and verb that can stand alone is called an *independent clause* (or *main clause*). An independent clause can be a simple sentence. An independent clause also can be the main clause in a sentence that contains another clause which is dependent on the main clause (it is dependent because it isn't a complete thought). An example of a dependent clause would be "because I am an industrial engineer." It's a clause—it's got a subject (*I*) and a verb (*am*)—but it cannot stand alone because of the word *because*. A dependent clause is also called a *subordinate clause*.

A *coordinating conjunction* is the name given to any of five main connectors (*and, or, but, nor, for*). The coordinating conjunction ties together or "coordinates" things of equal value. If it ties together two main clauses, the result is a compound sentence.

Now, with an understanding of these few terms, you're ready to add the appropriate punctuation.

The Comma

Comma marks the shortest pause required in breathing, a concept dating back to the days of oratory. The problem, of course, is that many people still punctuate according to their own need for air. When they stop to breathe, they stick in a comma. Many times, that's the right place for a comma. Many other times, it's not. The following few, simple rules can give you a lot of leverage in making decisions about the comma based on something other than the need for a breath of air.

Use a comma in the following cases:

1. *To separate two main clauses joined by a coordinating conjunction*

> *Example:* I like learning new things, but the rules for commas are too hard to learn.

Note: If you haven't got two main clauses—no matter how long or short the sentence—you don't need the comma. The most common error is in a sentence like this: "Grammar is not impossible to grasp and can be mastered with a little practice." The sentence has a coordinating conjunction (*and*) but it has only one main clause. The subject (*grammar*) has two verbs (*is* and *can be mastered*), so you've got a single subject with a compound verb. That adds up to a simple sentence. No comma needed.

2. *To separate introductory material from the main clause*

> *Examples:* Starting on the first of the month, employees must take no longer than thirty minutes for lunch.
>
> After that, they have to return to work.
>
> If this new rule goes into effect, I'll quit.
>
> Yesterday, I took a long lunch hour.
>
> First, I think I'll talk to my boss.
>
> When we have a chance to sit down together, I'm sure we can work something out.

Note: Grammar books will break my catch-all phrase "introductory material" into all kinds of wonderful grammatical terms like participial phrases, prepositional phrases, dependent adverbial clauses, and so forth; but the term "introductory material," no matter what kind, covers all that, and the rule works: whatever comes before the main clause is separated from the main clause by a comma. Even though some grammarians have eased up on the use of the comma in some cases, you'll never be wrong in applying this rule or any of the other rules laid out here.

3. *To separate items in a series*

> *Example:* John went fishing, hiking, and boating this past summer.

Note: "Items" can be anything—words, phrases, even main clauses. If they're in a series—three or more—use commas to separate them. The "old school" says to put a comma before the last one in the series, too. I still recommend that procedure because it enhances clarity. Otherwise, there could be confusion over whether the last two items are supposed to be treated as two or one. For example, "We will be moving the parts, production, material and storage departments." Does the writer mean three or four departments? If you normally put the comma in for all items in the series, the reader will know that when you leave it out, you mean three departments, not four. By the way, if you used *and* to connect each item in the series, you would not use commas to separate the items since *and* takes the place of the commas.

> *Example:* We will be moving the parts and production and material departments.

4. *To separate interrupters*

> *Example:* We believe, on the other hand, that the merger will be approved.
>
> Ben Smith, who is sitting in the back of the room, is of this opinion.
>
> The alternatives are not, however, within the scope of this discussion.

Here again, the catchword "interrupters" covers a lot of territory. Interrupters can be nonrestrictive phrases and clauses, conjunctive adverbs, words out of the proper order, and anything else that breaks or *interrupts* the normal flow of the sentence from subject to verb to object.

Notice that interrupters require commas on both sides. The commas, in effect, throw a fence around the interrupter, isolating it from the rest of the sentence. Another way of identifying an interrupter is to see whether you can lift it out of the sentence and still retain the meaning. If you can, you've got an interrupter.

> *Note:* If, however, the expression is essential to the meaning of the sentence, so that removing it would change the intended meaning, it is not an interrupter. In such cases, you have what's called a restrictive phrase or clause because the group of words limits or *restricts* the meaning of the sentence. That is, it is *essential* to the meaning of the sentence; therefore, it is not separated by commas.

> *Example:* All boys who are over six feet tall must play basketball.

If I put commas around the clause *who are over six feet tall,* I should be able to remove it from the sentence without losing or changing the sense of the sentence. Here, the removal of the clause would completely change the meaning of the sentence, so the clause is not an interrupter.

Another example: The store that carries computer equipment is on Main Street.

Big hint: Any clause introduced by the word *that* is restrictive, which means
that it is not an interrupter.

5. *To separate coordinate adjectives*

Coordinate adjectives work together to modify a noun or pronoun.

Example: The big, expensive-looking, beautiful car is owned
by the mayor.

Big hint: If you can stick in the word *and* in place of the commas and the
sentence still makes sense, you've got coordinate adjectives. Also, if you can
reverse the order of the adjectives without changing the logic of the sentence, you've got coordinate adjectives. In the example given, you could say,
"the beautiful, big, expensive-looking car," or "the big and expensive-looking
and beautiful car."

Note: Do not use a comma to separate adjectives if one adjective comes
before another adjective that is part of a complete unit of thought. In the example below, *old brick building* forms a unit of thought, so there is no comma
between *big* and *old.*

Example: The big old brick building is still standing.

6. *To separate participial phrases at the end of the sentence from the main
clause.*
Participial phrases are those expressions introduced by a participle, the
-ing forms of a verb, like *going* or *having* or *visiting.*

Example: Janet worked all summer, being too busy to take
a vacation.

Note: Almost everything else that comes at the end of the sentence does
not require a comma. Prepositional phrases, for example, do not.

7. *To separate items within dates and addresses*

Examples: June 5, 1987, is the release date.

My address is 102 South Main Street, Atlanta,
Georgia 30000.

Atlanta, Georgia, is my home.

Note: Don't forget the comma after the last item in a date or an address. Most people leave it out, but conventions require it. However, you do not put a comma between the state and zip code.

If you list only the month and year, you don't need a comma between them.

> *Example:* July 1986.

Military style places the day in front of the date and removes the commas.

> *Example:* 5 June 1985.

8. *To separate titles and other attachments after a person's or a company's name*

> *Examples:* The Bucket Company, Inc., is doing a booming business.
>
> John Jones, Jr., is addressing the group.

The Semicolon

With the mastery of commas behind you, you are ready for the even less well-known territory of the semicolon. Take courage. Contrary to what you might expect, the rules for the semicolon are really quite simple.

Use the semicolon in the following cases:

1. *To separate two main clauses joined without a coordinating conjunction*

The semicolon indicates a close connection between the ideas expressed in the two clauses, but the clauses must be separated.

> *Example:* I like my job; it has many rewards.

2. *To separate two main clauses joined by a conjunctive adverb (however, therefore, thus, nevertheless, consequently, etc.) or a transitional expression (on the other hand, in effect, as a result, etc.)*

> *Example:* I like my job; however, I want to move to another city.

Note: While there are only five coordinating conjunctions, there are many conjunctive adverbs and transitional expressions. If a conjunctive adverb or transitional expression separates two main clauses, the adverb or expression takes the semicolon *before* and the comma *after* it. Most people who make an error in this kind of construction use a comma to separate the clauses, thus producing a comma splice, the enemy of the English teacher! Remember also that a word like *however* can be an interrupter, requiring a comma on both sides. As with all of these rules, you have to know the way in which the word is being used in the sentence to choose the correct punctuation in each case.

3. *To separate items in a series when the items contain internal punctuation*

> *Example:*   The company does business in Atlanta, Georgia; Knoxville, Tennessee; and Greensboro, North Carolina.

In this case, the semicolons indicate the breaks that would ordinarily be indicated by commas, but since commas are used within the items in the series, the semicolons show the separation of the items themselves.

4. *To separate two main clauses joined by a coordinating conjunction when one or both of the clauses have internal punctuation.*

> *Example:*   John, who has studied for weeks for the bar exam, cramming every night, had a headache the day he took the exam; and he's not sure, given his state of health, his nervousness, and his lack of sleep, whether he passed.

Like the semicolons in the example for rule #3, this semicolon indicates the "big break" in the sentence, which the comma usually shows. Since the clauses themselves have commas within them, the semicolon shows the main division between the clauses.

Note: Any internal punctuation within a main clause permits the writer to use a semicolon between main clauses; however, most writers don't apply the rule unless the semicolon is needed for clarity.

The semicolon can be a powerful tool in writing. It's impressive when used correctly. And it's easy to use with confidence since there are only a few rules to master.

The Colon

Generally more readily understood as an organ of the body than a mark of punctuation, the colon is really not so mysterious. A few simple rules will shed much light on this often-misunderstood mark of punctuation.

1. The colon generally means *the following is/are.*

2. The colon follows a complete thought.

3. The colon never follows a form of the *to be* verb (*is, was, will be,* etc.).

4. The colon never follows a preposition (*to, from, by, over,* etc.).

> *Examples:* He has several good traits: honesty, dependability, and sincerity.
>
> There is one thing you must never forget: accuracy.
>
> *Incorrect usage:* I want to be: rich, famous, and happy. [See #3 above.]
>
> I am fond of: reading, writing, and arithmetic. [See #4 above.]

To correct the errors in the preceding examples, just take the colon out, and you have sentences with items in a series. If you understand the first two rules of the colon, rules #3 and #4 are just excess baggage.

Note: In much technical and business writing, the rule of the colon is often ignored with lists. This does not, however, make it right.

The colon, of course, has lots of other uses generally understood by all:

Use:	*Example:*
in time	10:15 AM
in headings	*Summary:*
in salutations	Dear Mr. Smith:
	Dear John:

Dashes, Parentheses, Square Brackets

Because of the similarity of their functions, dashes, parentheses, and square brackets are seldom used correctly in formal writing. A brief explanation of their correct use follows.

Dashes represent a major break in thought or an afterthought. They are also used to indicate items in a series in the middle of a sentence.

> *Examples:* There are three—make that four—programs to consider.
>
> The main points to consider—cost, effectiveness, maintenance, and return on investment—are discussed in this report.

Note: A dash is formed by typing two hyphens together (--) with no space on either side.

Parentheses, like dashes, indicate a break in thought or an afterthought. They can also be used to indicate additional information or a clarification or an acronym and to insert numbers.

> *Examples:* He considered the odds (having no other choice) and decided against the purchase.
>
> Metropolitan Atlanta Rapid Transit (MARTA) is scheduled for completion in September.
>
> We require a deposit of $65 (sixty-five dollars).
>
> We must decide whether to (1) ship the order, (2) deliver the order ourselves, or (3) refuse to fill the order.

Square Brackets indicate explanations or changes within a quoted passage or parentheses within parentheses.

> *Examples:* "The evaluator [Timothy Thompson] is correct in his assessment."
>
> "He presented his findings to [Jane Johnson] at the board meeting."
>
> "She suggested that he read several sources, among them, *Prose and Cons: The Do's and Don'ts of Technical and Business Writing* (a recent book on the style of reports [1986])."
>
> "They asked to recieve [*sic*] the information in triplicate."

Note: Sic is a Latin term used to indicate an error in the material being quoted.

Quotation Marks

Quotation marks have simple rules concerning punctuation.

1. In American usage, the period and comma *always* go *inside* the closing quotation marks.

> *Example:* He is a "Type A," working all the time.
>
> She said, "It's really a very simple rule."
>
> He called her behavior "uncharacteristic."

2. The semicolon and colon *always* go *outside* the closing quotation marks.

> *Example:* Let's talk about "gobbledygook": the jargon that inhibits communication.

3. The question mark and exclamation point go *inside* the closing quotation marks if they're part of the passage quoted; *outside,* if they're not.

> *Examples:* I have read the article "Are You Listening?"
>
> Have you read the article "Listening Made Easy"?

Note: Punctuation never doubles up at the end of a sentence, so one mark of punctuation serves the whole sentence. In the first example above, you wouldn't put a period after the question mark.

11
The Grammar Grind

We've been over the nuts and bolts of the style and technique of technical and business writing. It seems appropriate to finish with the grammatical odds and ends, the little things that probably won't keep you from getting a promotion or making yourself understood but that, if mastered, will elevate your writing to new levels of clarity and correctness.

Since textbooks have been written, and continue to be written, on the meaty aspects of grammar, I won't go into the sort of detail you can find in any good grammar handbook. What follows is a discussion of some of the trickier points of everyday grammatical usage.

Subject-Verb Agreement

1. *And* and *Or:* If "John and Mary *are* here" and "John or Mary *is* here" (both of these constructions are correct), then what do you do with this next one? "John or the twins *are/is* here." You make the verb agree with the subject closer to it. So, "John or the twins *are* here." Inversely, "The twins or John *is* here." The same would hold true for the questions, "*Are* the twins or John here?" and "*Is* John or the twins here?"

2. *Neither-Nor* and *Either-Or:* The rule for *or* applies as well to *neither-nor* and *either-or.* The verb agrees with the subject closest to it.

> *Examples:* Either the Smiths or Bill Johnson *is* attending the meeting.
>
> Neither Bill Johnson nor the Smiths *are* attending the meeting.

3. *Intervening Expressions:* In the sentence "The captain and the co-captain command the plane," the subject is compound (two subjects joined by *and* to make a plural subject). But what do you do with the following sentence? "The captain along with the co-captain *command/commands* the plane." In this case, the subject—*captain*—is singular and the group of words *along with the co-captain* is an intervening expression which does not affect the subject. So the verb—*commands*—must be singular. Intervening expressions include *along with, in addition to, together with, accompanied by, no less than, not to mention, including,* and *as well as.*

4. *Collective nouns:* Collective nouns include many people or things but are thought of as a single unit. Therefore, they take a singular verb.

> *Examples:* Management is concerned about rising production costs.
>
> The management team reports to the president.
>
> The committee meets every Tuesday.
>
> The company feels that its employees should receive Christmas bonuses.

Some grammarians say that collective nouns can be singular or plural, depending on the writer's desire to express the noun as a unit or to refer to the individuals within the unit. To eliminate confusion, I recommend always treating collective nouns as singular. To describe the individuals within the unit, change the construction from *team* to *members of the team* or from *management* to *managers.*

Inverted Word Order

Sentences traditionally read subject-verb-object. We've come to expect that arrangement. So when the normal order changes, many of us have trouble locating the subject to make it agree with the verb. For example, if I say

"Here are the ways I want this problem solved," I have to know that the subject will be plural before I get to it since it comes after the verb. (The subject is *ways,* not *here.*)

You may think my example simple; the mind is easily capable of anticipating the subject. But what do you do with the following sentence? "Here *is/are* the report and memo you wanted." If you chose the verb *are,* you'd be right because the subjects are *report* and *memo.* The word *and* connecting them makes them plural, even though each is singular by itself. Of course, the best way to avoid such dilemmas is to avoid the *here-there* construction since it's generally wordier than need be, and it keeps you from knowing the subject until after you read the verb.

Indefinite Pronouns

Indefinite pronouns include *each, either, neither, one, everyone, someone, no one, anybody, somebody, everybody.* They're easy to remember because you can add the words *one* or *single* to the indefinite pronouns to remind yourself that they're singular.

> *Examples:* Each (one), every (single) body, every (single) one.
>
> *Everybody has* a job that *he likes.*
>
> *No one wants his* job taken from *him.*

Now, these examples point up a problem with the indefinite pronoun. Since it is singular, it takes a singular verb and a singular pronoun to agree with it. And since it's *indefinite,* it stands for no one in particular. But, the singular pronoun is *he* or *his,* which carries a masculine connotation, even though it's supposed to be generic, that is, standing for all.

In an attempt to eliminate the problem, linguists and grammarians have suggested several solutions:

1. You can change the indefinite singular pronoun to a definite plural noun. Instead of saying, "Everybody has a job he likes," say, "All employees have a job they like." (*They* doesn't convey masculine or feminine; it could be either or both.)

2. You can use *he or she* and *his or her* whenever you need a personal pronoun.

> *Example:* Everybody has a job he or she likes.

3. You can use *he* or *his* one time; *she* or *her* the next.

Examples: Everybody has a job he likes.

Nobody wants to lose her job.

As you can see, there are problems with all three alternatives. The first solution forces you to use the plural when you may want to describe the individual.

In the second solution you may get tangled up in the repetition of *his or her* and *he or she* over and over again. While I've occasionally used this solution in this book, it poses problems since it makes the sentences longer and is often awkward.

In the third solution, you risk confusing your reader by switching from masculine to feminine in different sentences or paragraphs. The reader may wonder whether you specifically mean men one time and women another. This alternative works best when you have several examples you can use to illustrate your point. In one, you can use feminine pronouns; in another, masculine.

Other possibilities include the following:

1. Eliminate the pronoun.

   Example: Each employee must show a time card (eliminates the problem of *his* time card).

2. Repeat the noun instead of using a pronoun.

   Example: If the worker loses the time card, the worker must consult the payroll department.

Finally, however, what has come into widespread use is not among any of the alternatives listed above. It is also not grammatically correct. The practice avoids the problem of the masculine pronoun by changing the singular possessive pronoun to a plural.

   Example: Everybody likes *their* job.

Thus far, grammarians have banded together to disallow this one. But I suspect the popular vote will eventually win out. Until that time, however, you need to choose from among the acceptable alternatives to avoid offending readers of either sex.

Some indefinite pronouns, such as *none, some, part, all, half,* take a singular or plural verb, depending on the meaning of the sentence. When they refer to a singular word, the verb is singular. When they refer to a plural word, the verb is plural.

Examples: Some of the money is available.

Some of the coins are valuable.

None of the money has been spent.

None of the negotiations have proved fruitful.

Pronoun Agreement

In addition to the problem of pronoun agreement with indefinite pronouns, sticky points arise concerning pronoun agreement with nouns. Basically, the pronoun rules follow the subject-verb agreement rules. As in the discussion of subject-verb agreement, whenever a singular verb is required for a singular subject, the appropriate pronouns are also singular.

Examples: The committee elects its new officers.

John or the twins have their tickets.

The twins or John has his ticket.

One common error in pronoun agreement is with a word like *company* or *management.* Since each is a singular noun, each requires a singular pronoun for correct agreement.

Examples: The company has set its goals for the next fiscal year. (*Not:* The company has set their goals.)

Management plans to review its long-range goals. It will report to the company in June. (*Not:* Management plans to review their long-range goals. They will report to the company in June.)

Pronoun Reference

A pronoun needs to refer specifically to something in a sentence. The thing referred to is called the *antecedent* because it comes before the pronoun. Problems arise with pronouns like *this, that, they,* and *you* when the pronouns don't have a specific antecedent.

This or *that* shouldn't refer to a whole sentence or idea. Each pronoun should have a specific reference.

> *Example:* Some people can't sleep at night worrying about their work. This becomes a problem.

What does *this* refer to specifically? To clear up the problem, link the pronoun with the specific noun you have in mind, and you eliminate the unclear pronoun reference.

> *Correction:* Some people can't sleep at night worrying about their work. This lack of sleep becomes a problem.

They is often used as a general reference to no one in particular.

> *Example:* The government is responsible for the economy. They raise taxes.

They isn't anyone, and the pronoun doesn't point to anything. The correct pronoun should be *it*—*it* [the government] *raises taxes*—or another noun like *Congress.*

You produces similar problems, as the following example shows.

> *Example:* New employees should read the company style manual. In studying the document, you can learn a lot about company procedures.

If the writer is directly addressing the reader, *you* is the correct pronoun. But, most often, the writer is addressing a general readership, and *you* is a shift in address, indicating that the writer has suddenly turned from addressing a third person general reader to second person direct address.

> *Correction:* In studying the document, the new employee can learn a lot about company procedures.

Who and Whom

Who and whom present particular problems. Use *who* in the *subject* position and *whom* in the *object* position in the sentence.

> *Examples:* Who is there?
>
> I am the one *who* called.
>
> I am the one to *whom* you spoke.

I am the one *who* is to address the meeting.

The man *who* is fishing is my neighbor.

With *whom* did you discuss this matter?

Whom did they elect to the board?

Who can be the subject and *whom* can be the object of a sentence (independent clause), or *who* can be the subject and *whom* can be the object of a modifier (dependent clause). In each case, you base your decision about *who* (subject) or *whom* (object) on the use of the word in the *clause* in which it is located, not on the way in which the clause is used. In the sentence "I remember who the culprit is," the dependent clause *who the culprit is* serves as the direct object of the main clause *I remember;* but the decision about the word *who* is not based on the fact that the clause is used as a direct object in the sentence, only on the way in which the word is used in its clause. In this case, the pronoun required is the subject pronoun, *who.* There are trickier examples than this one, but the rule remains the same.

Note: After the pronoun *who* or *whom,* the verb in the clause agrees with the antecedent of the pronoun.

> *Examples:* He was the employee who was able to solve the problem. (*Who* refers to *employee*, so the verb *was* agrees with *employee.*)
>
> He was one of the employees who were able to solve the problem. (*Who* refers to *employees*, so the verb *were* agrees with *employees.*)
>
> *Exception:* In the expression *the only one,* the verb and any other pronouns in the clause introduced by *who* are always singular.
>
> He was the only one of the employees who was able to solve the problem. (*Who* refers to *the only one*, so the verb *was* agrees with *only one.*)

Reflexive and Intensive Pronouns

Reflexive and intensive pronouns are the "self" pronouns—*myself, himself, herself, ourselves, yourselves,* etc.

"Self" pronouns can have only two uses, neither of which is to take the place of a personal pronoun like *I, me, you, he, she, us,* etc.

In *reflexive* usage, the pronoun *reflects* or refers to the noun or pronoun preceding it.

> *Examples:* I saved myself five dollars.
>
> John brought fame and fortune upon himself.

In *intensive* usage, the pronoun *intensifies* or *emphasizes* the use of the pronoun or noun that precedes it.

> *Examples:* I myself did the job.
>
> Mary did it herself.

It is incorrect to say "John and *myself* will attend the meeting" or "On behalf of John and *myself,* we thank you" or "Just between John and *myself,* I didn't think the meeting was well run" or "He gave it to John and *myself.*" In each of these examples, the pronoun is *replacing* another pronoun, not *reflecting* or *intensifying* another pronoun.

People often want to use *myself* instead of the appropriate pronoun because they aren't sure whether the sentence should be "On behalf of John and (*me* or *I*), we thank you" or "He gave it to John and (*I* or *me*)." In both instances, the correct pronoun is *me*, the object of the preposition. Another reason that people use *myself* (and all the other "selves") incorrectly is that they think it sounds more intelligent and sophisticated. It doesn't.

Possessives

The rule for possessives is very simple. Yet there seems to be a lot of confusion about it.

If a word ends in *s*, add an apostrophe to make it possessive. If it does not end in *s*, add *'s*.

Now, there are screwy exceptions for proper nouns (names), but no one seems to agree on them. For instance, some authorities say that with proper nouns, if the noun is one syllable and ends in *s*, add an *'s*, as in John *Jones's* home. Others say to add *'s* with words of two or more syllables, as in Bill *Burgess's* home, but not with one-syllable words. Still others say to stick to the standard rule of the possessive no matter how many syllables the proper noun has. Since this standard rule is acceptable, I recommend following it in all cases.

Another problem arises when people don't use the apostrophe with possessives, as in the sentence "*Todays* society has become permissive." Or, even worse, people use the apostrophe when they mean plural (and this I see more and more), as in the sentence "The burger's are ready to eat."

Just remember that the possessive indicates ownership. When you write "the dog's tail," what you're saying is "the tail of the dog." If you mean more than one dog, you want to make the word plural *first*, then make it possessive. If the plural has an *s* on the end of it, add an apostrophe. If it doesn't have an *s*, add *'s*.

> *Examples:* The dog's tail
>
> The child's book
>
> The children's books
>
> The glass' color
>
> The glasses' color

Then there is the possessive pronoun *its*. And it has lots of relatives like *hers, his, ours, yours,* etc. These are called possessive pronouns because they already show possession. Therefore, they don't need an apostrophe.

What complicates matters, of course, is that the apostrophe can also indicate contractions like *it's* for *it is* or *it has*. If this usage confuses you, then do not use contractions. If you eliminate the use of the contraction *it's*, you know the possessive has to be *its*. If you want the plural possessive it's not *its'* (there's no such thing), but *their*.

> *Examples:* The dog wagged its bushy tail.
>
> The dogs wagged their bushy tails.

And that's it.

Suggested R and R (Reading and Reference)

Books

Bernstein, Theodore. *The Careful Writer.* New York: Atheneum, 1982.

Brusaw, Charles T., Gerald J. Alred, and Walter E. Oliu. *Handbook of Technical Writing,* 2d ed. New York: St. Martin's, 1982.

The Chicago Manual of Style, 13th ed. rev. Chicago: University of Chicago Press, 1982.

Follett, Wilson. *Modern American Usage.* Edited and completed by Jacques Barzun. New York: Hill and Wang, 1966.

Fowler, H.W. *A Dictionary of Modern English Usage,* 2d ed. rev. by Sir George E. Gowers. New York: Oxford University Press, 1965.

Gordon, Karen Elizabeth. *The Well-Tempered Sentence: A Punctuation Handbook For the Innocent, the Eager, and the Doomed.* New York: Ticknor & Fields, 1983.

_____ . *The Transitive Vampire: A Handbook of Grammar for the Innocent, the Eager, and the Doomed.* New York: Times Books, 1984.

Harrington, John, and Michael Wenzl. *A Suitable Design: How to Organize Your Writing.* New York: Macmillan, 1984.

Hayakawa, S.I. *Language and Thought in Action,* 4th ed. New York: Harcourt Brace Jovanovich, 1978.

Houp, Kenneth W., and Thomas E. Pearsall. *Reporting Technical Information,* 5th ed. New York: Macmillan, 1984.

Lanham, Richard A. *Revising Business Prose.* New York: Charles Scribner's Sons, 1981.

Mathes, J.C., and Dwight W. Stevenson. *Designing Technical Reports.* Indianapolis: Bobbs-Merrill, 1976.

McGraw Hill Dictionary of Scientific and Technical Terms, 2d ed. New York, 1978.

Michaelson, Herbert B. *How to Write and Publish Engineering Papers and Reports.* Philadelphia: ISI Press, 1982.

MLA Handbook. New York: Modern Language Association, 1984.

Newman, Edwin. *A Civil Tongue.* Indianapolis: Bobbs-Merrill, 1975.

_____ . *Strictly Speaking: Will America be the Death of English?* Indianapolis: Bobbs-Merrill, 1974.

Safire, William. *On Language.* New York: Times Books, 1981.

Sparrow, W. Keats, and Donald H. Cunningham. *The Practical Craft: Readings for Business and Technical Writers.* Boston: Houghton Mifflin, 1978.

Strunk, William, Jr., and E.B. White. *The Elements of Style,* 3d ed. New York: Macmillan, 1979.

U.S. Government Printing Office Style Manual, rev. ed. Washington, D.C.: Government Printing Office, 1967.

Journals

Bulletin of the Association for Business Communication

IEEE (Institute of Electrical and Electronic Engineers) Transactions on Professional Communication

Journal of Business Communication

Journal of Technical Writing and Communication

Technical Communication

Exercises

1. Getting Started

The exercises in this book are designed to allow you to use your own material and to stimulate thinking about good writing. For this reason, not every question is answered. Where helpful, some suggestions are included. Answers and suggestions follow these exercises.

Try these brainstorming exercises.

1. Consider the following hypothetical situation. Your boss has called you into his office. When you arrive he is frantically stuffing papers into his brief-case. "Fly ash!" he says. "Find me everything you can on fly ash!" He snaps the case shut and rushes toward the door. "Have it on my desk when I get back from Philadelphia on Tuesday. Can't stop to talk any more. I'm late for my plane." He leaves abruptly. You can't get in touch with him to clarify the assignment. Look up *fly ash* in a dictionary, but don't do any other research yet. Instead, use brainstorming to try to figure out what your boss wants on his desk Tuesday.

Note: Of course, you won't be certain you are right on this one, but brain-storming may help you come up with something. If you are really stuck, look at some of the suggestions on p. 141.

_____

_____

_____

_____

_____

_____

2. You have to spend a week in Chicago on business. The need came up suddenly, and the trip cannot be postponed. Your assistant must fill in for you while you are gone. Use brainstorming to list what you need to tell him or her.

Note: If you are a student, try brainstorming a list for someone who is filling in for you while you are off campus for a couple of days. That person may be attending classes for you and/or handling other, extracurricular responsibilities.

_____

_____

_____

_____

_____

_____

To Outline or Not to Outline?

3. Whether produced from an outline or not, good technical writing is logically organized and can therefore be outlined after it is written. If you receive written reports, take one from your files. [If you are a student, obtain a report from your instructor.] Go through it, reading only the first sentence of each paragraph as well as any headings. Does the report make logical sense when read this way? If it is a good report, this kind of reading will often be like reading a summary. Go through one of your own reports the same way.

4. Outline a report on your activities for the past week. Divide your activities into at least four main headings and have at least two subheadings under each heading. If you are working full time, imagine that your boss has asked that you give him or her a report on your on-the-job activities. If you are a college student, imagine that an instructor has asked you to outline a report on your class work and your extracurricular activities for the past week. [In either case, exclude your love life.]

_____

_____

_____

_____

_____

_____

_____

_____

_____

_____

_____

_____

How about a Handbook?

Here is a tip for choosing a good handbook:

5. Look up two or three points you are certain about. Does the handbook have these points right? [I always check to see what the handbook says about *it's. It's* is a contraction for *it is* or *it has. Some handbooks don't mention *it has.*]

6. Look up two or three points that have always confused you a little (or a lot). Is the handbook's explanation clear?

If the answer to either of these questions is no, keep looking.

2. How Do You Like My Style?

1. Take a page of your own writing and go through it carefully, circling each form of the verb *to be (be, is, am, are, was, were, being, been)*. Now try to rewrite the whole page so that you don't use any of these forms. Try to make the page read so smoothly that no one will notice anything peculiar about the writing.

2. Calculate the Gunning Fog Index for the paragraph beginning right under the heading PARAGRAPHS on page 9 of this chapter. Now calculate the Fog Index for a 100 + word passage from your own writing. Are you assuming a higher or lower grade level for your readers than the author of this book?

3. Using the transitional expressions as clues, rearrange the sentences below into a logical, unified paragraph:

However, I am currently engaged in a critical phase of the Danforth recycling project.

Please have these figures on my desk by Friday, June 7.

Therefore, I must delegate the job to you or a member of your department.

Elliot Edam, Vice-President in Charge of Development, has asked me to compile figures on cheese imports from Gorgonzola between 1979 and 1984.

_____

_____

_____

_____

_____

_____

_____

4. Here are some old sayings and quotations translated into gobbledygook. See if you can ungobble them.

a. Optimal timely seam repair can reduce garment downtime by 90%.

_____

b. Continuance of existence or its alternative is the issue of deliberation.

_____

c. Issuance of promissory notes and receipt thereof are activities of questionable merit.

_____

d. Persons of limited intellectual capacity move from a state of financial stability to a state of financial instability within a short period of time.

_____

5. The Following sentences contain unnecessary passive-voice verbs. Rephrase them with active-voice verbs.

a. The contract will be signed by our agent as soon as the confirmation is sent by your company.

_____

_____

b. The Universal Law of Gravitation was formulated by Isaac Newton after an apple was observed falling from a tree.

_____

_____

c. You will be notified if any other information is needed.

_____

_____

d. Your letter has been received and your request will be given top priority.

_____

_____

e. It was reported today by the *New York Tribune* that stupidity may be contagious.

_____

_____

6. Some of the following sentences use personal pronouns where they should not. Others avoid using those personal pronouns unnecessarily. Rephrase the sentences properly.

a. After the wire has been attached, the screw should be tightened carefully.

_____

_____

b. I feel the facts indicate we should abandon this project.

_____

_____

c. A sucker should never be given an even break by the reader.

_____

_____

d. The writer has been asked to study the feasibility of marketing a bag-lady doll to supplement our line of fashion dolls.

_____

_____

e. This venture should, I think, return a net profit on investment of 32% during the first year.

_____

_____

3. Know Your Audience

1. Below are the audience types described on pp. 28–29. Beside each type write either a) the name and title of one person in your organization (other than yourself) who belongs in that category or b) the name of a group or department which consists of people in that category.

General reader(s): _____

Technician(s): _____

Engineer(s) or technologist(s): _____

Specialist(s): _____

Manager(s): _____

User(s): _____

Note: If you are a full-time student, try using your school as the organization.

2. To which of the above do you most frequently address memo reports? Is the person (or the group) a peer? _____

3. From the company files, pull a memo report with headings and sub-headings. [If you are a student, obtain such a report from your instructor.] In the margin beside each heading, write the type(s) of reader(s) (from the list in Question 1) who might be expected to be interested in that section. [You may not, of course, find a section for every type of reader.]

4. If you have recently written a two- or three-page report, either at work or as a class assignment, note in the margin the type of reader most likely to be interested in each headed section (or, if you did not head the sections, each paragraph). How many potential types of readers does your report have? _____

5. Suppose you work for Biscomb's Better Biscuits, which produces "the best biscuit baked between Boston and Berkeley." You have just learned

of a new commercial baking sheet called Potter's Perfectly Perforated Precision Pan, which cuts baking time by 50 percent and which your company might want to use in its bakery. Below are four readers for a memo report you plan to write. Beneath each name and description, write one question which must be answered for that person in your report.

Puffem Mupp, Chief Baker, graduate of Cupcake Cooking College, oversees the baking operation. His position is much like plant foreman in a factory.

_____

_____

Clickcoin P. Penney, Vice-President in Charge of Finance, has a B.S. in economics and an MBA. He is also a CPA and is in charge of all the financial aspects of the company.

_____

_____

Bertram Biscomb, president of the company, has only a high-school education, started the company from scratch after losing his job during a recession. Biscomb is practical and businesslike and attributes his success to producing a superior product at a fair price.

_____

_____

Nutson Boltz, Chief Engineer at Biscomb's, has a Bachelor of Science degree in industrial engineering technology. He is in charge of the baking plant operation, including all aspects of its operation from receipt of materials to shipment of the finished goods.

_____

_____

4. Let's Get Organized

1. Refer to Question 5 at the end of the preceding chapter. Create at least one heading for each reader of the report on Potter's Perfectly Perforated Precision Pans. [Some readers may, of course, read the whole report, but your sections in this exercise should each be designed with the needs of one particular reader in mind.]

2. From which reader—Biscomb, Mupp, Penney, or Boltz—are you most likely to meet resistance to Potter's Perfectly Perforated Precision Pans in each of the following situations?

Situation 1: The use of the pans will necessitate reorganizing the traffic pattern of the plant.

Situation 2: The use of the pans will mean a change in the formula for Biscomb's Biscuits.

Situation 3: The use of the pans will mean retraining the bakery workers.

Situation 4: The pans will not pay for themselves unless sales can be doubled.

3. From your company files, take a report with several levels of headings. [If you are a student, obtain such a report from your instructor.] Write the headings down in order, indenting the second-order headings half an inch, the third-order headings a full inch, and so on. Do you get a logical outline?

4. Rewrite the following paragraph using bullets:

Using Potter's Perfectly Perforated Precision Pans will increase our production by 50%, prevent overbaking of biscuits near the edge, cut energy costs by approximately 3%, and cut absenteeism among workers due to heat prostration by approximately 5%.

5. From your files, take a memo report at least two pages long. [Students, obtain one from your instructor.] Rewrite the first paragraph so that it acts as a summary which answers all of the following questions for the primary reader(s):

- What problem or need does this memo address?
- What has been done, is being done, or needs to be done about this problem or need?
- What is the purpose of the report itself?
- What action does the report require or recommend that the reader(s) take as a result of reading the report?

5. Graphics Are Grabbers

1. Below is a paragraph from a completely incompetent technical report on two tests which measure executive potential. (The report was written by George Flubb, who has no potential but who has relatives in high places.) DON'T try to rewrite this hopeless mess. Instead, use the data to make a table so that you (and your reader) can easily compare the two tests.

Executive Tests, Inc., has one of those seminar rooms and the meeting was held in one. Harold Jones—he is a vice-predsient—talked about these two tests his companies got that tests people to see if they would make good executives. The first test he talked about I don't think Id like to take. It makes you do some writting and junk like that about whoat youd do if such and so happened. (jones said that was a hypothetical decision-making situation whatever that means.) Writting isnt my strong point I guess you can tell. (Ha, ha!) Anyway, there's this other test they call the Executive Index Test. (The first one is the decision Analysis test.) He said this Execative index Test has multiple-choice questions (about 150) and cost $50. I guess its cheaper because its graded by amachine and the other one has to be graded by a person because of the writting part and the other one'z more expensive because somebody's got to grade all that writting and junk. Instead of a machine. Anyway, they charge $125 a throw for that one and it tells you where a guy who takes it will probly wind up in mangement and gives you his strong points and weak points. The other one does that to. Some of the people had questions but we had to leavve early because Harrys wife was sick not because of the ballgrame (ha ha). So we came on back and got home about 2:15 after stopping for lunch, which costed us $6.73 including the tip and the round trip gas mileage was 106 miles. Oh, yes, we can get the grade back on the multiple choice test in a week. The other one takes longer.

2. In each of the following cases, decide which kind of graph (line, bar, or circle) would be best to display the data.

a. The nation of Gorgonzola exports four kinds of cheeses—Edam, colby, cheddar, and mozzarella. In 1985, 2964 tons of cheese were exported: 397

tons of Edam, 433 tons of colby, 1755 tons of cheddar, and 379 tons of mozzarella.   ans: _____

b. The average per capita net income of Gorgonzolan cheesemongers shows an interesting trend:

1980	$ 6,956
1981	7,660
1982	8,433
1983	8,367
1984	9,675
1985	10,758

ans: _____

c. The dollar value of Gorgonzola's five major exports in 1985 is as follows:

Cheeses	$205,874
Caskets	173,979
Pickled pigs' feet	136,908
Instant baked Alaska	128,565
Sports cars*	54,900

*Two exported

ans: _____

6. Editing: The Garbage Collector

In all of the exercises below, try to eliminate wordiness whenever possible.

1. Revise the following sentences for impact by using more specific, concrete nouns and verbs.

a. The table was of a fibrous material.

_____

b. It is necessary to have a level surface to work on.

_____

c. The material has poor organization.

_____

2. Revise the following sentences to strengthen the beginnings.

a. There are some problems connected with the adoption of Potter's Perfectly Perforated Precision Pans. For example, the formula for Biscomb's Better Biscuits would have to be altered, and the traffic pattern in the plant would have to be changed.

b. It is expected that the changeover will take approximately three days.

c. There is a reason for this. The Potter Pans are slightly wider than our current pans.

3. Revise the following sentences to eliminate pointless repetition. (Consult the lists on p. 59 if necessary.)

a. For the past history of this operation, refer back to p. 106 of my memo of August 12, 1985.

b. Potter's Perfectly Perforated Precision Pans are rectangular in shape and have surfaces which are black in color.

c. Our final conclusion is that the desirable benefits are sufficient to outweigh the undesirable drawbacks.

4. Revise the following sentences to reduce or eliminate the wordiness.

a. In the year 1985, we had a total output in the neighborhood of 200 million biscuits and 7.6 billion crumbs.

b. We conducted an investigation of waste and reached the conclusion that, in all probability, we should give consideration to packaging and marketing the crumbs as Biscomb's Better Buttercrumb Breading.

---

c. In addition to the above, the results of our investigation led us to the conclusion that it is necessary for us to cut our costs by an amount equal to something like $256,000 per year.

---

5. Revise the following sentences by substituting simple, direct words for the inflated language.

a. The initial step is to implement the changeover to Potter's Pans. Subsequent to this changeover, the crumb marketing operation should be initiated.

---

b. It is not feasible to utilize crumbs as insulation in domestic habitats.

---

6. Revise the following sentences by using concrete, specific words in place of the vague language. (Don't be afraid to guess at the writer's meaning. Your object is to come up with something more specific. Everyone won't choose the same specifics.)

a. Application of manufacturing by-products to organizational environments is one parameter for future analysis.

---

b. The device most applicable in this area will function as an interface.

---

7. Revise the following sentences to eliminate the dangling modifiers.

a. Changing to Potter's Pans, waste will be cut by 25%.

---

b. After beginning our modification, our competitors may do likewise.

---

8. Revise the following sentences to reduce the size of the noun clusters. If you can eliminate a cluster, do so.

a. Enclosed is our biscuit crumb factory feasibility study report.

---

b. Consult the factory pan changeover operation procedures information manual for further details.

---

9. Revise the following sentences to reduce or eliminate the prepositional-phrase strings.

a. Our study of procedures in factories for biscuit baking on a large scale is inconclusive.

---

b. We expect an increase in profits with a decrease in waste from use of new procedures.

---

10. Revise the following sentence to make the structure parallel.

Our company will benefit because waste will be cut, new markets, and increasing production.

---

7. Cases in Point

Read, try to decipher, then rewrite for clarity each of the following cases. Most are from actual sources.

1. Copernicus believed that the earth went around the sun but conven-

tional astronomers didn't. (From an informal history of astronomy.)

2. The shutter is holding open as the button is being depressed. Releasing the button will finish the shutter. (From the instruction book for a camera manufactured abroad.)

3. Proffessor Simonton is to tuff. I made a A or B in everthing else but he give me a F in Eglish. (From a student letter protesting a grade.)

4. Proper and humane social behavior toward other people may be determined by assessing one's reaction to such behavior as manifested in others toward oneself. (A well-known quotation translated into inflated language.)

5. We have added ten new emergency vehicles to our fleet. This has been a life-threatening benefit to the community. (From a county safety commission report.)

6. We had to drag the river for the body. The swimmer had dived off a cliff and hit his head on a rock. This took about ten hours. (From the same safety commission report.)

7. Random events will selectively occur at times when said events can minimize optimum outcome. (A familiar "scientific proverb" translated into inflated language.)

No amount of rewriting will correct the problem in the following set of instructions. First try to figure out what it says. Then, figure out what the problem is.

8. The timer button is located in the upper right-hand corner. Holding it in, press down the spring-loaded release lever in the upper left-hand corner. At the same time, turn the intensity control knob (bottom middle of the panel) to 1.8. (From instructions for an unspecified device.)

8. The Proof of the Writing Is in the Reading

Proofread the passage below for the following kinds of mistakes:

• spelling

• hyphenation

• forms of numbers

• capitalization

• numerical accuracy

Note: The text has been doublespaced for your convenience. You may wish to photocopy the page and make corrections on the copy. To correct misspelled words, simply strike through the error and write the correction ABOVE the line.

PROPORSAL FOR CONVERTING BAKIRY

TO POTTER RECISION PANS

Our baking opertion has proved reasonably efficient in the past, but recently an imporved baking pan, Potter's Perfectly Perforated Presiscion Pan, was developed. Consequently, I was asked to study the feasibilty of using this new pan. My findings, detailed in the report below, indicate that the Potter Pan can increase our production by 60%.

Background

For the past seven years our bakry has been producing at full capacity, approximately 200-million biscits per year. Althogh market reserch indicates that could sale 300 million, it has not been feasable to ncrease our production because buying new ovens and packaging equipment would double our capacity to 400-million per year. However, it is unlikely that we could take advantage of this increased production by doubling our sales any time in the near future.

Thus, it would be vest to find some way of increasing our production by about fifty percent without making major equipment purchases. Potter Precsion Products claims that its new Pan can increase production for a bakery such as ours by approximately 60%. Since such an increase would allow us to meet the potential demand for our prodict, we decided to study the feasibility of using the Potter Pan in our bakery.

Cost

The changeover will take approximately 3 days and will require modification of the bakery's thirty eight twelve rack ovens. The estimated total cost is as follows:

475 Potter Pans @ $15 each	$7500.00
Modification of ovens	1520.00
Total	$9070.00

Production will cease durning the three day changeover. However, employies will recieve full salary during this peroid. Increasded production after the changeover should offset this cost. If the corportion choses to borrow money for the changeover, interest charges should be added to the total.

9. Diction's Dirty Dozen

Circle the correct word or phrase in each case.

1. One drunken basketball player can affect/effect the whole team's morale.

2. The riot at the Wednesday Evening Tea and Bridge Club had two affects/effects: serious injury to the players and cancellation of the tournament.

3. Even if something affects/effects you, it does not affect/effect a real change unless its affect/effect is permanent.

4. The rock star's many fans divided his clothing among/between themselves.

5. The new board game *Real Life* forces players to choose among/between two undesirable alternatives.

6. One can judge the popularity of any event by the amount/number of cars illegally parked nearby.

7. The new Secaucus Low Food Diet requires the dieter to consume fewer/less calories and to do fewer/less exercise than does the Nevada Cream Cheese Diet.

8. No matter how bad/badly things seem, you need not feel bad/badly unless you did your job bad/badly.

9. The Bill of Rights comprises/is comprised of the first ten amendments to the United States Constitution.

10. His confusion seemed to be because of/due to a lack of common sense.

11. Because of/Due to this peculiarity, he was elected President. (After all, most voters are confused, too.)

12. We cannot tunnel any farther/further until we have farther/further information.

13. These kind/kinds of wridwights are always used with this type/types of blickstine.

14. The President is an amiable-type/amiable type of/amiable person.

15. The lead/led in a pencil is actually graphite. What leads/leds us to believe otherwise? Lead/Led pipe hasn't been used for years either; galvanized iron and copper are used instead.

16. A leader who has never lead/led anything may mislead/misled his followers.

17. We are laying/lying on the grass. A duck waddles up and lays/lies an egg on Abner's nose. Moral: Never lay/lie near a duck. You may raise/rise up with egg on your face.

18. Also, never set/sit eggs on a chair where you may set/sit.

19. Egg placement raises/rises serious questions.

20. Through a hole in your pocket you may loose/lose all your loose/lose change. You cannot afford such a lose/loss.

21. The school principal/principle had one principal/principle in dealing with unruly students.

22. A poor person may have principals/principles but no money. A rich person, even one without morals, may draw interest on his or her principal/principle.

23. An office supply store which sells stationary/stationery may not remain stationary/stationery during a hurricane.

24. Do I think her behavior is coarse/course? Of coarse/course I do! She could give a coarse/course in coarseness/courseness to sandpaper.

10. Punctuation Pointers

Using the eight rules for commas, insert commas where needed in the following sentences. (Remember: Always use a comma to separate introductory material from the main clause.)

1. We shall meet next on January 10, 1988 at the Merchantsville Town Hall 888 Avocado Avenue Merchantsville California 91212.

2. Our speaker will be J. Callaway McBurger Ph.D.

3. We may however have to postpone our meeting if the roof leaks; good weather on the other hand does not guarantee a good meeting.

4. Dr. McBurger will speak on three matters of concern: agriculture pre-Columbian culture and yogurt culture.

5. Although planning this far in advance is often dangerous we expect a good turnout.

6. After we had seen films of tidal waves earthquakes and volcanic eruptions did not seem so bad. (Be careful with this one.)

7. A firm well-prepared foundation is essential if we hope to erect large residential buildings.

8. The crew quit work at ten o'clock feeling that nothing more could be accomplished.

9. Everyone was interested in the subject of the speech but the speaker was so dull that only three people stayed to the end.

10. Because we have six members who insist on paying their dues on time the Procrastination Club is calling a disciplinary meeting which will be held next year or the year after to consider how this infraction of the rules should be handled.

In each of the following pairs of sentences, one of the two correctly uses punctuation while the other does not. Put a check mark beside the correct sentence.

1. **a.** Jack Handle, the club president, Myron Upp, the secretary, and Cliff Hanger, the treasurer, met last week.

   b. Jack Handle, the club president; Myron Upp, the secretary; and Cliff Hanger, the treasurer, met last week.

2. **a.** The officers of the club are also the only members, therefore, they can complain only to themselves.

   b. The officers of the club are also the only members; therefore, they can complain only to themselves.

3. **a.** We have four main problems: corruption, apathy, carelessness, and catatonia.

   b. Our four main problems are: corruption, apathy, carelessness, and catatonia.

4. **a.** Our officers, Jack Handle, Myron Upp, and Cliff Hanger, cannot handle these problems alone.

   b. Our officers—Jack Handle, Myron Upp, and Cliff Hanger—cannot handle these problems alone.

5. **a.** We must increase membership, lower dues, and have more frequent meetings, and if those measures do not work we must consider disbanding.

   b. We must increase membership, lower dues, and have more frequent meetings; and if those measures do not work, we must consider disbanding.

6. **a.** During the campaign, Phillip Pole announced "My worthy opponent (he meant Chatsworth Longtalk) believes that government should tax the very air we breathe".

   b. During the campaign, Phillip Pole announced, "My worthy opponent [Chatsworth Longtalk] believes that government should tax the very air we breathe."

7. **a.** "I don't believe in that", Pole continued. "After all, wouldn't the cost of metering be prohibitive"?

   b. "I don't believe in that," Pole continued. "After all, wouldn't the cost of metering be prohibitive?"

8. **a.** Was it not Patrick Henry who said, "Give me liberty or give me death?"

   b. Was it not Patrick Henry who said, "Give me liberty or give me death"?

9. **a.** The National Aeronautics and Space Administration (NASA) hopes to fund its space-shuttle flights partly through contracts with private enterprise.

   b. The National Aeronautics and Space Administration [NASA] hopes to fund its space-shuttle flights partly through contracts with private enterprise.

10. **a.** NASA itself seeks the services of such companies and institutions as: Boeing Aircraft, California Institute of Technology, and the Smithsonian Institution.

   b. NASA itself seeks the services of such companies and institutions as Boeing Aircraft, California Institute of Technology, and the Smithsonian Institution.

11. The Grammar Grind

Circle the correct word choice in each case.

1. Neither the bricks nor the paved slab is/are adequate in this case.

2. Both he and she has/have volunteered for the potato-tasting project.

3. Music, in addition to the other arts, pays/pay little until the creator is dead.

4. The audience at a rock concert often goes/go wild; sometimes the performing group does/do, too.

5. There is/are frequently very good reasons for apparently stupid acts. Careless mistakes, as well as negligence, is/are sometimes encouraged by managers who wish to appear more competent than their subordinates.

In the following sentence, try to eliminate the problem with the indefinite pronoun by using each of the methods described on pp. 109–111. Finally, select the sentence that seems best to you.

6. Biscomb's Better Biscuit Company wants every one of its customers to be satisfied with their purchase.

a. _____

b. _____

c. _____

d. _____

e. _____

In each of the following sentences, circle the choice that makes the pronoun agree with its antecedent (the word the pronoun refers to).

7. Next week, the Supreme Court will determine its/their position on liability of crumbcake manufacturers.

8. Customers of Ace Crumbcake Enterprises have filed a class action lawsuit against the company because it uses/they use eraser crumbs as an ingredient.

9. The customers/group believe that the outcome of their case will determine the future of crumbcake in the United States.

10. Congress has a crumbcake bill on its/their agenda for the next session.

11. Crumbcake lovers everywhere hope that the Supreme Court in its/their wisdom will decide in its/their favor.

Correct the pronoun reference in each of the following sentences by striking through the error and writing the correct choice above the line.

12. A first offender may be put into a cell with hardened criminals. What you learn in prison often turns you into a hardened criminal, too.

13. The officers objected to the change in the retirement system, but the board refused to reinstate the old rules and also turned down a request for a "grandfather clause." This is what caused the demonstration.

Circle the correct choice, *who* or *whom*, in each of the following sentences.

14. Who/Whom did you want?

15. I never know who/whom to blame.

16. The Society for the Preservation of Seedy Districts presented its award to the member who/whom had been most derelict during the year.

17. I don't know who/whom is in charge.

Circle the correct choice in each of the following.

18. Charles and I/me/myself welcome you to the meeting of Birdwatchers Anonymous.

19. By compulsively watching birds, I harm no one but I/me/myself.

20. The organization presented the Bronze Bird to Amos and I/me/myself.

Insert apostrophes where necessary in the following sentences.

21. Its almost certain our dog can find its way home.

22. Bill and Tom Gander, John Ganders sons, were over at the Drakes house last night; the boys were talking to Susan Drake.

23. Two weeks salary is too much to spend on ones girlfriends birthday present. One weeks salary is plenty.

24. Four weeks from now we can expect the committee to submit its reports.

25. Mens fashions do not change as fast as womens.

Answers to Exercises

1. Getting Started

1. *Suggestions:* Your boss might want

   a pile of fly ash

   a paragraph defining fly ash

   a bibliography of recent articles on fly ash

   Note: You may be able to think of additional possibilities. The first item on the list here is obviously absurd, but you should have some absurd things on a brainstorming list. Giving your boss a combination of the second and third items, however, might be a good solution to a difficult problem.

2. How Do You Like My Style?

2. The Gunning Fog Index for the paragraph is 12. The author is assuming twelfth-grade reading level. Only 11% of the words are "hard words," but the average sentence length is 19 words.

3. Elliot Edam, Vice-President in Charge of Development, has asked me to compile figures on cheese imports from Gorgonzola between 1979 and 1984. However, I am currently engaged in a critical phase of the Danforth recycling project. Therefore, I must delegate the job to you

or a member of your department. Please have these figures on my desk by Friday, June 7.

4. **a.** A stitch in time saves nine.

   b. To be, or not to be; that is the question.

   c. Neither a borrower nor a lender be.

   d. A fool and his money are soon parted.

5. Suggested rewrites:

   a. Our agent will sign the contract as soon as your company sends the confirmation.

   b. Isaac Newton formulated the Universal Law of Gravitation after observing an apple fall from a tree.

   c. We will notify you if we need any other information.

   d. We have received your letter and will give your request top priority.

   OR, BETTER STILL,

   We will give your request top priority. [How would you know about the request if you hadn't received the letter.]

   e. Stupidity may be contagious, according to today's *New York Tribune.*

6. **a.** After attaching the wire, tighten the screw carefully.

   b. The facts indicate that we should abandon this project.

   c. Never give a sucker an even break.

   d. I have been asked to study the feasibility of marketing a bag-lady doll to supplement our line of fashion dolls.

   e. This venture should return a net profit on investment of 32% during the first year.

3. Know Your Audience

5. Here is a list of possible questions. You may have come up with others which are even better.

Puffem Mupp: Will the workers' duties be substantially the same if the Potter pans are installed?

Clickcoin P. Penney: How much do the Potter pans cost?

Bertram Biscomb: If we use the Potter pans, will the biscuits produced still be the same quality?

Nutson Boltz: Will changing to Potter pans call for changes in the shipping operation?

4. Let's Get Organized

2. *Situation 1:* Nutson Boltz

   Situation 2: Bertram Biscomb

   Situation 3: Puffem Mupp

   Situation 4: Clickcoin P. Penney

4. Using Potter's Perfectly Perforated Precision Pans will do all of the following:

   • increase our production by 50%

   • prevent overbaking of biscuits near the edge

   • cut energy costs by approximately 3%

   • cut absenteeism among workers due to heat prostration by approximately 5%

5. Graphics Are Grabbers

1.

Name of Test	Type of Questions	Method of Grading	Results Available in	Cost per Person Tested
Executive Index Test	Multiple choice	machine	one week	$ 50
Decision Analysis Test	Essay	hand	?*	$125

*George Flubb's report does not give a turnaround time for the Decision Analysis Test.

2. **a.** A pie, or circle, graph is probably the best choice since the cheeses listed represent 100% of a whole.

   b. A line graph is best for showing any trend.

   c. A bar graph is the best choice here. A pie graph is inappropriate because the listed exports are not the *only* exports.

6. Editing: The Garbage Collector

1. Suggested rewrites for specificity and concreteness:

   a. The worktable was made of plywood.

   b. Do the work on a level counter or tabletop.

   c. The proposal needs separate sections on scope and methodology.

2. Here are some suggested revisions:

   a. The adoption of Potter's Perfectly Perforated Precision Pans will cause some problems: for example, we will have to alter the biscuit formula and change the traffic flow in the plant.

   b. The changeover will take approximately three days.

   c. . . . because the Potter pans are slightly wider than our current pans. (Combine this clause with another sentence.)

3. Suggested revisions:

   a. For the history of this operation, see p. 2 [if you eliminated needless repetition there, too] of my August 12, 1985, memo.

   b. Potter's Perfectly Perforated Precision Pans are rectangular and have black surfaces.

   c. The benefits outweigh the drawbacks.

4. Suggested revisions:

   a. In 1985, we had a total output of approximately 200 million biscuits and 7.6 billion crumbs.

    b. Our findings suggest that we should consider packaging and marketing the crumbs as Biscomb's Better Buttercrumb Breading.

    c. In addition, we must cut our costs by approximately $256,000 per year.

5. Suggested revisions:

    a. The first step is to install Potter's pans. Next, the crumb marketing operation should begin.

    b. Using crumbs as insulation in homes is not feasible.

6. Suggested revisions:

    a. We should study the feasibility of using crumbs in office-building construction.

    b. The modem will allow us to use the word processor as a terminal with the mainframe computer.

7. Suggested revisions:

    a. Changing to Potter's pans will cut waste by 25%.

    b. After we begin our modification, our competitors may do likewise.

8. Suggested revisions:

    a. Enclosed is our feasibility report on biscuit-crumb manufacture.

    b. For further details, consult the manual on pan changeover.

9. Suggested revisions:

    a. Our study of large-scale biscuit baking is inconclusive.

    b. Using these new procedures, we can expect higher profits and less waste.

10. Suggested revision:

Our company will benefit from waste reduction, new markets, and increased production.

7. Cases in Point

Suggested revisions are given below.

1. Unlike conventional astronomers of his time, Copernicus believed that the earth went around the sun.

2. Depress the button to open the shutter.
Release the button to close the shutter.

3. (Not much can be done with this one except to correct the grammar and spelling and maybe refine it a little.)

Professor Simonton is too tough. I made an *A* or a *B* in every one of my other courses, but I got an *F* in Professor Simonton's English class.

4. Do unto others as you would have them do unto you.

5. (The first sentence is all right. Change the second.)

Suggested change:

Thanks to this new equipment we can now handle life-threatening emergencies more effectively.

6. (Drop the first sentence. Keep the second. Revise the third to include information from the first sentence.) Suggested revision:

After dragging the river for about ten hours, we finally recovered the body.

7. If anything can go wrong, it will. (Murphy's Law)

8. The problem with the last exercise item is, of course, that the instructions can be followed only by someone with three hands.

8. The Proof of the Writing Is In the Reading

PROPOSAL   BAKERY

~~PROPORSAL~~ FOR CONVERTING ~~BAKIRY~~

POTTER'S PRECISION

TO ~~POTTER RECISION~~ PANS

operation
Our baking ~~opertion~~ has proved reasonably efficient in the past, but re-

improved   Precision
cently an ~~imporved~~ baking pan, Potter's Perfectly Perforated ~~Presiscion~~ Pan,

feasibility
was developed. Consequently, I was asked to study the ~~feasibilty~~ of using

this new pan. My findings, detailed in the report below, indicate that the Pot-

ter Pan can increase our production by 60%.

Background

bakery
For the past seven years our ~~bakry~~ has been producing at full capacity,

200 million biscuits   Although   research
approximately ~~200 million bisoits~~ per year. ~~Althogh~~ market ~~reserch~~ indicates

we   sell   feasible   increase
that could ~~sale~~ 300 million, it has not been ~~feasable~~ to ~~ncrease~~ our produc-

tion because buying new ovens and packaging equipment would double our

400 million
capacity to ~~400 million~~ per year. However, it is unlikely that we could take

advantage of this increased production by doubling our sales any time in

the near future.

best
Thus, it would be ~~vest~~ to find some way of increasing our production by

50%   Precision
about ~~fifty percent~~ without making major equipment purchases. Potter ~~Pree-~~

pan
~~sion~~ Products claims that its new ~~Pan~~ can increase production for a bakery

such as ours by approximately 60%. Since such an increase would allow

product
us to meet the potential demand for our ~~prodict~~, we decided to study the

feasibility of using the Potter Pan in our bakery.

Cost

three
The changeover will take approximately ~~3~~ days and will require modifi-

38   twelve-rack
cation of the bakery's ~~thirty eight twelve-rack~~ ovens. The estimated total cost

is as follows:

	$7125.00
475 Potter Pans @ $15 each	~~$7500.00~~
Modification of ovens	1520.00
	$8645.00
Total	~~$9070.00~~

Production will cease ~~durning~~ the ~~three-day~~ changeover. However, ~~employies~~ will ~~recieve~~ full salary during this ~~peroid.~~ ~~Increasded~~ production after the changeover should offset this cost. If the ~~corportion choses~~ to borrow money

(corrections above lines) during — three-day — employees — receive — period Increased — corporation chooses

for the changeover, interest charges should be added to the total.

9. Diction's Dirty Dozen

The correct word choices in each sentence are listed below. Where the sentence contains two problems, the correct choices are listed in order.

1. affect
2. effects
3. affects
   effect
   effect
4. among
5. between
6. number
7. fewer
   less
8. bad
   bad
   badly
9. comprises
10. due to
11. because of
12. farther
    further
13. kinds
    type
14. amiable
15. lead
    leads
    Lead
16. led
    mislead
17. lying
    lays
    lie
    rise
18. set
    sit
19. raises
20. lose
    loose
    loss
21. principal
    principle
22. principles
    principal
23. stationery
    stationary
24. coarse
    course
    course
    coarseness

10. Punctuation Pointers

The commas are used correctly in the sentences below.

1. We shall meet next on January 10, 1988, at the Merchantsville Town Hall, 888 Avocado Avenue, Merchantsville, California 91212.

2. Our speaker will be J. Callaway McBurger, Ph.D.

3. We may, however, have to postpone our meeting if the roof leaks; good weather, on the other hand, does not guarantee a good meeting.

4. Dr. McBurger will speak on three matters of concern: agriculture, pre-Columbian culture, and yogurt culture.

5. Although planning this far in advance is often dangerous, we expect a good turnout.

6. After we had seen films of tidal waves, earthquakes and volcanic eruptions did not seem so bad.

7. A firm, well-prepared foundation is essential if we hope to erect large residential buildings.

8. The crew quit work at ten o'clock, feeling that nothing more could be accomplished.

9. Everyone was interested in the subject of the speech, but the speaker was so dull that only three people stayed to the end.

10. Because we have six members who insist on paying their dues on time, the Procrastination Club is calling a disciplinary meeting, which will be held next year or the year after, to consider how this infraction of the rules should be handled.

The correct choice, a or b, is given below beside the number of the sentence:

1. b	6. b
2. b	7. b
3. a	8. b
4. b	9. a
5. b	10. b

11. The Grammar Grind

The correct choices are as follows:

1. is

2. have

3. pays

4. goes, does

5. are, are

6. Answers will vary.

The correct choices for pronoun-antecedent agreement are the following:

7. its

8. it uses

9. customers

10. its

11. its, their

Here are some possible corrections. Some variation should be expected.

12. First offenders may be put into cells with hardened criminals. What those first offenders learn in prison often turns them into hardened criminals, too.

13. The officers objected to the change in the retirement system, but the board refused to reinstate the old rules and also turned down a request for a "grandfather clause." The denial of the "grandfather clause" caused the demonstration.

The correct choice, *who* or *whom*, is shown below.

14. Whom

15. whom

16. who

17. who

The correct choice is shown below.

18. I

19. myself

20. me

The corrected sentences are shown below.

21. It's almost certain our dog can find its way home.

22. Bill and Tom Gander, John Gander's sons, were over at the Drake's house last night; the boys were talking to Susan Drake.

23. Two weeks' salary is too much to spend on one's girlfriend's birthday present. One week's salary is plenty.

24. Four weeks from now we can expect the committee to submit its reports. (There should be no apostrophes in the sentence.)

25. Men's fashions do not change as fast as women's.

Index

Abstracts, 42
Acronyms, 19, 86
Active voice, 20-21
Adjectives, 101
    coordinate, 101
    cumulative, 101
Adverbs, conjunctive, 102
Affect-effect, proper usage of, 89-90
Agreement, sentence structure, 107-108, 111
American National Standards Institute, 36
Among-between, proper usage of, 90
Amount-number, proper usage of, 90-91
Analogy, 22
Analysis, in writing for readers, 29
ANSI, 36
Apostrophe, 114-115
Arabic numbers, 2-3, 36, 85
Arrangement, in writing, 2-5, 9-10, 27, 31-45
Audience, 25-29, 33, 55
    and visual aids, 55
    attitude toward subject, 29
    engineers, 28
    general readers, 28
    managers, 29
    primary readers, 26
    secondary readers, 26-27
    specialists, 28
    technicians, 28
    technologists, 28
    users, 29
    what is read, 33

Bad-badly, proper usage of, 91
Because of-due to, proper usage of, 92
Between-among, proper usage of, 90
Brackets, square, use of, 104-106
Brainstorming, 2, 119-120, 141
Bullets, for organizing, 36-37
Bureaucratese, 11-16

Capitalization, 85-86
    abbreviations, 86

acronyms, 86
days of week, month, holidays, 85
directions, 86
proper names, 85
seasons, 86
titles, 85-86
Clauses, 98-100
    dependent, 98
    independent, 98
    main, 98-99
    nonrestrictive, 100
    subordinate, 98
Collective nouns, 108
Colon, use of, 103-104
Comma splice, 102-103
Comma, use of, 98-102
Comprise-consist, proper usage of, 91-92
Conjunctions, coordinating, 98
Connectors, 97
Consist-comprise, proper usage of, 91-92
Coordinating conjunctions, 98
Cut and paste, as organizing method, 3-5

Dangling modifiers, 62-63
Dashes, use of, 104-105
Deductive order, 10
Diction, 89-96
Dictionaries, 5, 82-83
Direct object, 20, 94, 113
Double numeration, as heading system, 36
Due to-because of, proper usage of, 92

Editing, 11-16, 45, 57-68
    concrete words, 62
    dangling modifiers, 62-63
    fence straddling, 66
    first drafts, 45
    inflated language, 11-16, 61-62, 66
    noun clusters, 63-64
    parallel structure, 64-65

prepositional phrases, 64
redundancies, 59
sentence construction, 58-59
sentence impact, 58-59
"so-what?" test, 66-67
tips and pointers, 58-65
wordiness, 60-61
Effect-affect, proper usage of, 89-90
Exercises, for writing self-help,
    119-151

Farther-further, proper usage of,
    92-93
Fence straddling, 66
Figures, 47-51, 86
Fog Index, 8-9
Forecasting statements, 10, 40-41
Further-farther, proper usage of,
    92-93

Generalization, 10
Gobbledygook, 11-16, 71-74
Grammar handbooks, 6
Graphics, 47-55
    figures, use of, 47-51
    graphs, use of, 49, 51-55
    tables, 47-51
    visual aids, 48-55
Graphs, 49, 51-55
    abscissa, 51, 52, 53
    bar graph, 49, 53-54
    coloring and shading, 53
    horizontal scale, 51
    line graph, 49, 51-53
    ordinate, 51, 53
    pie graph, 49, 53, 55
    scale, 51
Gunning Fog Index, 8-9

Headings, 33-36, 86
    double numeration system, 36
    first order, 34
    main headings, 34-36
    outline system, 36
    proofreading, 86
    second order, 34
    side, 34
    run-in, 34-35
    third order, 34
Hub and spokes, as organizing
    method, 5
Hyphenation, 82-84
    for clarity, 84
    numbers in combination, 83

two words as modifier, 83
    with prefixes, 84
Hypotheses, development of, 10

Imperative mood, 20
Indicative mood, 20
Inductive order, 10
Intensive pronouns, 113-114
Interrupters, 100
Intervening expression, 108
Introduction, 42
Introductory material, 99
Its, proper usage of, 115

Jargon, 19, 73-74

Kind-kinds, proper usage of, 93

Lay-lie, proper usage of, 94-95
Lead-led, proper usage of, 94
Legalese, 11-16
Lie-lay, proper usage of, 94-95
Lists, use of, 36-40
    bulleted, 36-37
    capitalization, 38-39
    numbered, 37-38
    parallel construction, 38-40
    punctuation, 38-40
Loose-lose, proper usage of, 95

Modifiers, dangling, 62-63
Mood, 20-21
    imperative, 20
    indicative, 20
    subjunctive, 20

Neologisms, 17-19
    -ity suffix, 19
    -ize suffix, 17-18
    -tion suffix, 18-19
Note cards, use of, 5
Noun clusters, 63-64
Nouns, collective, 108
Number-amount, proper usage of,
    90-91
Numbered lists, 37-38
Numbers, spelling of, 84

Officialese, 11-16
Organizing writing, 1-5, 9-10, 27,
    31-45
    abstracts, 42
    audience, 27
    brainstorming, 2

bullets, 36-37
characteristics, 7
cut and paste method, 3-4
editing, 45
forecasting statements, 40-41
headings, 33-36
hub and spokes method, 5
introduction, 42
lists, 36-38
note cards, 5
numbered lists, 37-38
ordering of parts, 32
organizational aids, 1-5
pyramid method, 5
reading habits, 33
S-curve, 5
starting, 1-5
summaries, 41-45
Outlining, 2-5
for audiences, 27

Paragraphs, 9-10
development of, 10
transitional words and expressions, 9-10
Parallel construction, 38-40, 64-65
in editing, 64-65
in lists, 38-40
Parentheses, use of, 104-105
Participial phrases, 101
Passive voice, 20-21
Personal pronouns, 21-22
Possessives, 114-115
Prefixes, 84
Prepositional phrases, 64
Principal-principle, proper usage of,
95-96
Pronouns, 21-22, 63, 109-114
agreement, 111
antecedents, 111-112
indefinite, 109-111
intensive, 113-114
it, use of, 63
personal, 21-22, 109
reference, 111-112
self, use of, 113-114
Proofreading, 81-87
Punctuation, 38-40, 97-106
colon, 103-104
comma, 98-102
dashes, 104-105
in lists, 38-40
parentheses, 104-105
quotation marks, 106

semicolon, 102-103
square brackets, 104-105
Pyramid, as organizing method, 5

Quotation marks, 106

Raise-rise, proper usage of, 94-95
Readability, 7-9
Fog Index, 8-9
formulas, 8-9
sentence, length of, 7-9
Readers, *see* Audience
Redundancies, 59
Reference guides, 5-6
dictionaries, 5
grammar handbooks, 6
thesauruses, 5-6
Reflexive pronouns, 113-114
Restrictive clauses and phrases, 100
Rise-raise, proper usage of, 94-95
Roman numerals, 2-3, 36

S-curve, as organizing method, 5
Semicolon, use of, 102-103
Sentence construction, 58-59
Sentences, 58-59, 97-98
complex, 98
compound, 97
compound-complex, 98
simple, 97
types of, 97-98
varying length and form of,
58-59
Set-sit, proper usage of, 94-95
Sic, definition of, 105
Sit-set, proper usage of, 94-95
"So-what?" test, 66-67
Square brackets, use of, 104-105
Stationary-stationery, proper usage
of, 96
Style, 7-23
analogy, use of, 22
bureaucratese, 11-12
characteristics of technical writing, 7
gobbledygook, 11, 73-74
jargon, 19, 73-74
legalese, 11, 12-16
neologisms, use of, 17-19
officialese, 11-12
paragraphs, 9-10
Plain Language Movement,
14-16
pronouns, personal, 21-22

readability, 7-9
sentence, length of, 7-9
style manuals, company, 23
voice, active, 20-21
Style manuals, 23
Subject-verb agreement, 107-108
and-or, 107
collective nouns, 108
either-or, 108
intervening expressions, 108
neither-nor, 108
Subjunctive mood, 20
Suffixes, 17-19
-ity, 19
-ize, 17-18
-tion, 18-19
Summaries, 41-45
abstracts, 42
introductions, 42

Tables, 47-51, 86
footnotes, 50
formal, 50
informal, 49-50
proofreading, 86
Thesauruses, 5-6
Topic sentences, 10
Transition, words and expressions,
9-10
for conclusions, 10
for differences, 9
for results, 10
for similarity, 9

for spatial order, 10
for step-by-step order, 9
Translations, from foreign language,
76-77
Type-types, proper usage of, 93

Verbs, 20-21, 91
linking, 91
passive, 20-21
transitive, 20
Visual aids, 47-55
figures, 47-49
flow charts, 53
graphs, 49, 51-55
labels, 49
lists of illustrations, 49
numbers, 49
organizational charts, 55
pictograms, 53
placement of, 48-49
schematic diagrams, 55
tables, 47-51
Vocabulary, choice of, 5-6, 17-19, 58-
62, 89-96
Voice, 20-21, 59
active, 20-21
passive, 20-21, 59

Westinghouse reading habit study,
33
Who-whom, proper usage of,
112-113
Wordiness, 60-61